Advances in Diptera - Insight, Challenges and Management Tools

Edited by Sarita Kumar

Published in London, United Kingdom

Advances in Diptera - Insight, Challenges and Management Tools
http://dx.doi.org/10.5772/intechopen.102160
Edited by Sarita Kumar

Contributors
Sarita Kumar, Arunima Sahgal, Haibin Han, Shujing Gao, Linbo Xu, Bo Zhang, Yanyan Li, Kejian Lin, Ning Wang, Wenyuan Niu, Wenchang Duan, Pervin Erdogan, Zemran Mustafa, Anita Solar, Aljaz Medic, Metka Hudina, Robert Veberic, Zeeshan Javed, Saira Mansha, Usama Saleem, Asad Mangat, Bilal Rasool, Muhammad Imran, Amna Batool, Mashal Shahzadi, Tehreem Raza, Muhammad Asrar

First published in London, United Kingdom, 2022 by IntechOpen
IntechOpen is the global imprint of INTECHOPEN LIMITED, registered in England and Wales, registration number: 11086078, 5 Princes Gate Court, London, SW7 2QJ, United Kingdom

British Library Cataloguing-in-Publication Data
A catalogue record for this book is available from the British Library

Additional hard and PDF copies can be obtained from orders@intechopen.com

Advances in Diptera - Insight, Challenges and Management Tools
Edited by Sarita Kumar
p. cm.
Print ISBN 978-1-80356-353-4
Online ISBN 978-1-80356-354-1
eBook (PDF) ISBN 978-1-80356-355-8

Meet the editor

Renowned entomologist Sarita Kumar is a professor at Acharya Narendra Dev College, University of Delhi, India. She has made an immense contribution to teaching, research and academic administration for over 29 years. Her work on insect pest and vector control includes 22 books on biological sciences, over 70 research publications, over 80 presentations/publications at national and international conferences, 10 e-book research chapters and 15 e-modules. She acts as editor/associate editor and reviewer for many journals of national and international repute. As a prominent curriculum/content developer, Prof. Kumar is recognized by renowned organizations as a key resource person. She has also received many prestigious Meritorious and Distinguished Teacher Awards.

Contents

Preface

Diptera, also called true flies, are soft-bodied, delicate, and usually small-sized insects. They comprise one of the major insect orders with rich species diversity. To date, more than 152,000 species of flies have been described within more than 188 recognized families. It is estimated that more than 1,000,000 species probably exist today that need to be explored and recognized.

Diptera are abundant throughout the world. They are reported to be prevalent in the tropics and sub-Arctic regions, at sea level, on beaches, and even on mountains. Only the cold Arctic and Antarctic ice caps are without these flies. Because of their ubiquitous presence in diverse habitats, dipteran flies are of considerable economic importance. They play a valuable and varied role in the ecosystem and human life. These tiny organisms are widely known for their role as model research organisms, ecological indicators, scavengers, pollinators, food for predators, and control agents of insect pests such as parasitoids and predators. The pestiferous groups can, however, also transmit serious ailments to humans and animals, and can significantly affect agriculture and forestry.

This book highlights advances in dipteran biology, their development, pest status and management. Conventional management strategies using chemical and biorational products, their bio-efficacy, and mode of action are covered, along with innovative and advanced control interventions in the form of mechanical devices, physiological disruptors, and those acting at biochemical and molecular levels.

The five chapters, arranged in three sections, each cover a different and novel aspect of the field of diptera. Section 1, 'Physiology of Diapause in Dipterans', covers research progress on the diapause of dipteran insects, a physiological process that enables them to survive in a natural environment otherwise unconducive to their survival. The phenomenon provides an adaptive advantage for flies, ensuring that their life cycle is synchronized for optimal growth, development, and reproduction. Diapause is categorized according to the requirement, the development stages of flies, and the steps in the process. Chapter 1 also elaborates on the environmental signals inducing and regulating diapause onset, sensitive insect states of diapause conditions, the endogenous molecular mechanism of diapause, and previous research results. The diapause process and intrinsic mechanisms in dipteran insects are analyzed in terms of endogenous changes in hormones, metabolism, genes and proteins, etc.

Chapters 2 and 3 under Section 2, 'Conventional and Advanced Strategies for Mosquito Management', focus on mosquitoes, the medically significant dipterans responsible for transmitting a number of serious diseases including dengue, Chikungunya and Zika viruses, malaria, encephalitis, filariasis and yellow fever. Chapter 2 illustrates the biology and diversity of the three main mosquito genera, *Anopheles, Aedes,* and *Culex*. Several traditional measures are discussed, such as use of biological agents – larvivorous fishes and crustaceans – and the application of chemical and plant-based toxicants in the larval habitat. Physical and behavioral processes, such as the frequent use of mosquito nets, repellents derived from plants, electrical traps and attractant traps in mosquito management, are described at length along with challenges in the implementation of these initiatives.

Chapter 3 discusses advances in mosquito control interventions devised to induce genetic, physical and behavioral modifications in mosquitoes. Strategies examined range from the sterile insect technique (SIT) inducing male sterility to reduce vector population, developing flightless females by the release of insects carrying a dominant lethal (RIDL), creating transgenics with abnormal and lethal genes, using gene drive technology to genetically modify mosquitoes and reduce their transmission competency, reducing the vectorial capacity by *Wolbachia* infection, and attracting and killing mosquitoes *via* application of attractive toxic sugar baits (ATSB). The use of lasers and light detectors to investigate mosquitoes' behavior, and enhance their trap and kill through high-power lasers are also described.

Section 3 is titled 'Biology and Management of Dipteran Pests'. Chapter 4 discusses dipteran pests that damage different parts of garlic and onion plants and cause extensive crop losses. The biology, life cycle, and pest status of the two most significant dipterans, *Delia platura* and *Delia antiqua,* are described, with cultural, biological and chemical control interventions for integrated management of these pests. The importance and management of minor pests such as leaf miners are also discussed.

Chapter 5 discusses the most significant walnut pest, the walnut husk fly (*Rhagoletis completa* Cresson), which causes crop losses of up to 80% in many growing regions, including the United States and Europe. The chapter describes the geographic distribution of this pest, its bionomics and infestation symptoms, and its effect on yield quantity and quality. New and traditional monitoring and control methods are reviewed; biotic protection and the possible role of phenolic compounds in cultivar resistance to walnut husk flies are emphasized.

I hope that this book will serve as a useful handbook for students, researchers and practitioners associated with the field of insects, and that the potential research directions it indicates will inspire future research ideas.

Sarita Kumar
Department of Zoology,
Acharya Narendra Dev College,
New Delhi, India

Section 1

Physiology of Diapause in Dipterans

Chapter 1

Research Progress on Diapause in Flies (Diptera)

Haibin Han, Yanyan Li, Bo Zhang, Kejian Lin, Shujing Gao, Linbo Xu, Ning Wang, Wenchang Duan and Wenyuan Niu

Abstract

Diapause is a physiological process in which insects can survive in a natural environment that is not conducive to their survival, which is the result of long-term adaptation to environmental conditions. It provides a great adaptive advantage for insects, allowing insects to survive in unsuitable seasonal environments to synchronize their life cycles with those suitable for growth, development, and reproduction. The process of regulating insect diapause is a complex process interacting with multiple mechanisms. In this chapter, a review is given of the current knowledge of diapause types, environmental inducing factors, sensitive states, and the endogenous molecular mechanism associated with diapause in flies (Diptera). Research regarding both the diapause process and intrinsic mechanism is reviewed.

Keywords: diapause, stages of diapause, biotic and abiotic factors, molecular mechanisms, Diptera

1. Introduction

Diapause is a state in which insects suspend or arrest the development in response to unfavorable environmental cues. It is an adaptive mechanism with a genetic basis, regulated by the external and internal environment factors, and occurs in a specific stage during the life cycle of an insect such as embryonic, larval, pupal, and active adult stage. Once induced, diapause cannot be immediately terminated even if unfavorable conditions disappear, unless a certain break period has been experienced. Diapause is terminated with the return of appropriate environmental conditions (temperature, light, moisture, etc.), and physical and chemical conditions [1]. Diapause provides an adaptive advantage for insects, allowing them to continue surviving in unfavorable seasonal environments and ensuring that their life cycle is synchronized with conditions suitable for growth, development, and reproduction.

There are two types of diapause, obligatory diapause and facultative diapause. Obligatory diapause, also known as absolute diapause, means that insects have to enter diapause to complete their life cycle, regardless of environmental conditions. It is most found in univoltine insects (one generation per year). For example, *Anthocharis cardamines* is a univoltine butterfly species that has an obligate pupal diapause in United Kingdom [2]. Facultative diapause, also known as random diapause, means that insects start this process only when environmental conditions become adverse. It is most commonly seen in bivoltine (two generations per year) or multivoltine insects (more than two generations per year). For example, the bivoltine strain of the silkworm, *Bombyx mori*, has a facultative embryonic diapause in Japan [3]. The generation of diapause is variable, but the life stage in which diapause occurs is fixed. Thus, photoperiod, together with temperature and other environmental factors, leads to a facultative diapause in insects. For insects, the significance of diapause is not only to enable them to survive the adverse environmental conditions, but also to make the population as uniform as possible, which greatly increases the possibility of male and female mating and thus ensures the reproduction of the population. To meet energetic costs, insects use two methods that are metabolic depression and energy storage. By reducing their metabolic consumption, insects can accelerate the accumulate energy to enhance their resistance to adverse environments [4]. In recent years, the research results on diapause of the order of Diptera have increased. This chapter classifies the research progress of diapause in Diptera, including the main groups, types, stages, parental effects, diapause-inducing factors, and mechanisms of molecular regulation. It provides theoretical support for population control of Diptera pests and effective biocontrol using natural enemies [5].

2. Taxa of diapause

Diptera is the fourth largest order after Coleoptera, Lepidoptera, and Hymenoptera. In terms of diapause research, 85 species of Diptera have been studied, including Tephritidae, Culicidae, Calliphoridae, Agromyzidae, Oestridae, Muscidae, Anthomyiidae, Tachinidae, Drosophilidae, and Cecidomyidae (**Figure 1**).

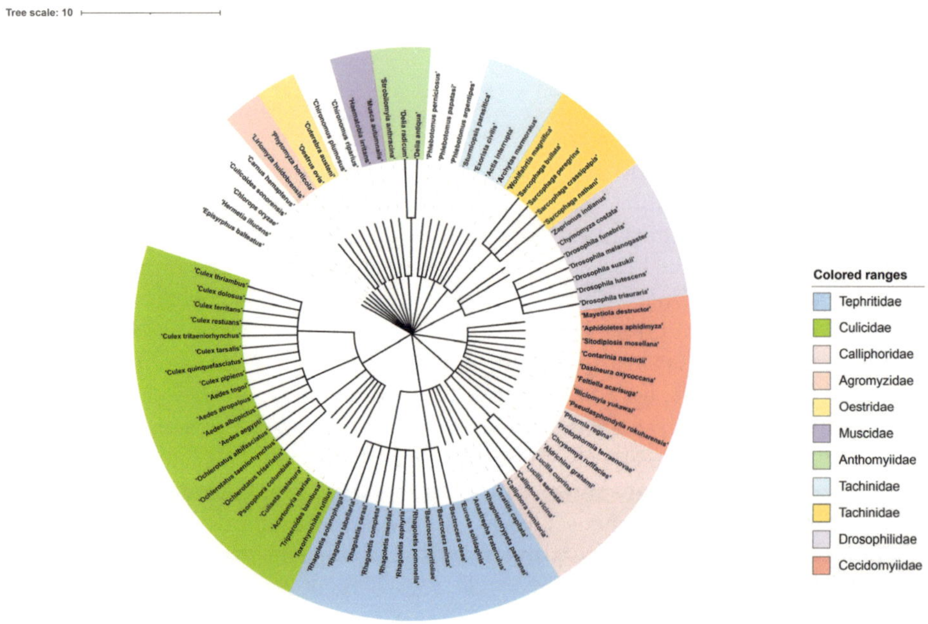

Figure 1.
Taxa of dipause in Diptera.

3. Types of diapause

Diapause can be categorized according to life stages as an egg (embryonic), larval, nymphal, or adult diapause (**Table 1**). Based on the hereditary feature, there are two types of diapause, one mandatory (obligate diapause) and one optional (facultative diapause) (**Table 1**).

Insects with obligate diapause enter diapause at a fixed developmental stage on schedule no matter how the external environmental conditions are, while those with facultative diapause can be induced into diapause at a certain stage but uncertain generation. Diapause is mainly affected by the changes in external environmental conditions. When the environmental conditions are unfavorable, insects enter the diapause, otherwise they continue to develop. Flies also experience diapause in the pupal stage (quiescent stage), during which the activity is extremely weak. *Sarcophaga similis* and *Haematobia irritans* have a pupal diapause [6], with the pupal diapause of the parasitoid fly *Exorista civilis* occurring within the insect host. Pupal diapause, common in other Diptera, is conspicuously absent in the Culicidae, whose members may pass through diapause as eggs, larvae, or adults. *Aedes albopictus* overwinters as diapausing eggs in temperate climates [7]. For *Chlorops oryzae*, diapause occurs as larvae, triggered by conditions experienced by the egg [8]. Diapause of *Procontarinia mangifera* is observed in larvae [9], Adult reproductive diapause occurs in *Drosophila suzukii* and *Protophormia terraenovae* [10]. The third-instar larvae of *Eurosta solidaginis* show obvious hibernation (winter diapause) characteristics [11]. Once *Lucilia sericata* reaches the third instar, larvae stop feeding and then enter diapause [12]. *Sitodiplosis mosellana* undergoes diapause as larvae in the soil [13].

Classification		**Characteristics**
Diapause induction stage	Embryonic diapause	occurs at any stage of insect embryo development; regulated by the brain-hypopharyngeal gland-diapause hormone endocrine system
	Larval diapause	occurs at any instar of larval larvae, but mostly occurs at the late larval stage
	Nymphal diapause	commonly happen in Diptera and Lepidoptera, which is regulated by brain-prothoracic-corpora allata interaction
	Adult diapause	commonly occurs in Coleoptera, Lepidoptera, Hemiptera, Diptera, Homoptera, and Orthoptera, and is regulated by the islet- corpora allata
Hereditary feature	Obligatory diapause	insects have no choice but to enter this process at some stage in their life.
	Facultative diapause	insects will start this process only when environmental conditions become adverse

Table 1.
Types of diapause.

4. Incidence of diapause

The diapause process can be divided into three phases: pre-diapause, diapause, and post-diapause. Pre-diapause occurs before the beginning of unfavorable environmental cues. Insects in this phase forecast an impending transformation in certain environmental stimuli in a special way, and change their internal neuro-endocrine system and metabolism level to enter diapause. At this stage, insects maintain normal development, which include induction and preparation stage [14]. The induction stage happens before the beginning of unfavorable environmental cues. Insects receive these specific environmental stimuli called "token stimuli" such as photoperiod, temperature, humidity, and food, to regulate their development and decide whether to enter diapause or not. Larvae of *S. mosellana* enter diapause at a long day length [15]. The preparation phase is followed by the induction phase, during which insects accumulate energy substances such as lipids, sugars, and amino acids. Lipids provide a large amount of nutrition for insects in diapause and prevent heat loss and mechanical damage [16]. Saccharides such as glycogen and trehalose, as instant sources of energy, play a crucial role in dealing with abiotic stresses [17]. Amino acids reserve help in both providing raw materials for protein synthesis of diapause insects and resistance to cold and desiccation. Amino acids are stored prior to diapause, but they help in maintaining life activities during diapause and post-diapause [16]. Previous studies of diapause in *Culex pipiens* demonstrated that during the diapause, mosquito accumulated more sugar than non-diapause mosquitoes [18].

Because of frequent observations of changing responses to various environmental conditions, diapause is divided into three eco-physiological sub-phases: initiation, maintenance, and termination [14]. The incidence of diapause is affected by a variety of factors. Many insects enter diapause at any stage of their life cycle, but for some species, the diapause stage is fixed, which can be judged by observing the color, appearance, and cocoon making of insect bodies [14, 19–21]. The maintenance phase refers to the period in which the insect remains undeveloped even under favorable developmental conditions, and the respiration and metabolic rates are at low levels. The diapause maintenance period of different insects varies greatly, ranging from a few weeks to several months or even several years. With the return of favorable environmental conditions, the intensity of diapause gradually decreases and enters the diapause termination. At this stage, insects are sensitive to temperature factors, especially low temperatures. In addition, photoperiod can induce diapause termination. Application of exogenous substances can also break insect diapause. For example, exogenous ecdysterone can terminate the diapause of *Bactrocera minax* [22].

In the termination phase, insects enter the next inactivity period if the environmental conditions are still unfavorable; however, they start resuming their physiological development [14].

5. Parental effects on diapause

Parents exhibit a greater effect on the diapause phenotype of their offspring, which defines as parental effect [23]. Parental effect is subject to natural selection, and it is the response mechanism of phenotypes to environmental heterogeneity [24, 25]. Among the parental effects, the female parent exhibits a greater influence than the male parent, so maternal inheritance is considered to be a short form of non-Mendelian parental effect (including maternal and paternal inheritance). For insects, female effects on offspring are relatively common. The parental effect of diapause in Diptera is mainly affected by environmental conditions such as diapause duration, photoperiod, temperature, and parental factors. Ordinarily, parents produce more diapausing progeny if they experience short-day length, limited resources, or low temperature [26]. In the study of *A. albopictus*, maternal photoperiod has a direct influence on egg size and embryogenesis [27].

Larval diapause in the blow fly *Calliphora vicina* is induced by their mothers to cope with short-day photoperiods. Due to the various photoperiod and temperature of the parents, the process of *C. vicina* larval diapause can vary in duration, indicating that the accumulations of diapause stimuli by parents significantly influence the intensity and incidence of progeny diapause [28]. The incidence of diapause is completely under maternal control and is not affected by the male [29]. *Lucilia sericata* has a facultative diapause in the third larval instar after cessation of feeding. Induction of the diapause is influenced by the photoperiod and temperature conditions experienced by insects in the parental generation as well as those experienced by the larvae themselves [30]. In *Sarcophaga bullata*, a maternal effect blocks the programming of diapause in progeny of females reared in short-day length that have experienced pupal diapause [31].

6. Environmental cues for diapause induction

Insect diapause is a complex process in which many mechanisms interact with each other. The major environmental signals regulating diapause onset in insects include temperature, photoperiod, food, and population density. Studies on insect diapause show that the primary cause and state of diapause can be identified and insects could be induced into diapause by artificially simulating diapause conditions in the field. Environmental cues, mostly temperature and photoperiod, control reproductive diapause in flies (Diptera), which regulate the induction, maintenance, and termination of diapause.

Seasonal change in photoperiod is the most reliable information to detect the time of year and is the major environmental signal regulating diapause onset in most insects. Photoperiod refers the period of time in a day that an organism is exposed to light or, more simply, day length. Photoperiodism is a biological response to a change in the proportions of light and dark in a daily (24 h) cycle, and the average number of daylight hours that cause insect diapause is known as "critical photoperiod." When insects respond to changes in light intensity through the brain or compound eyes, the internal "timer" automatically evaluates the length of day or night and regulates the insects to enter the diapause [32]. Usually, the diapause of Diptera is caused by short-day length. *A. albopictus* enters diapause with short-day length. The photoperiodic diapause is a crucial ecophysiological adaptation of *A. albopictus* to climate change in North America and strongly affects seasonal population dynamics, thus affecting the transmission potential of arboviruses [33]. *Chlorops oryzae* enters summer diapause under long day length, but the critical photoperiod is shortened with the increased temperature [34]. In *C. oryzae*, winter diapause in the first larval stage is induced by short-day length in the egg stage and maintained by short days in the larval stage [35]. *D. suzukii* is shown to be a typical of short-day length diapause species [9], and also, *S. similis* enters pupal diapause under short-day conditions [36].

Temperature is another major environmental signal regulating diapause, especially for Diptera. *Linothele sericata* can be stored at 7.5°C for several months and is seen as a potential replacement for honeybees, whose diapause is mainly affected by low temperatures in winter [37, 38]. The Chinese citrus fruit fly, *B. minax*, exhibits pupal diapause in the soil from November to March in the next year, and the pupal period increases with pupal weight. Temperature before entering diapause is a reasonable index to predict overwintering individuals [11]. *P. terraenovae* adults do not enter diapause at 30°C in either long-day or short-day conditions; however, between 17.5 and 27.5°C, the insect shows a long-day photoperiodic response, indicating that the diapause is mainly induced by low temperature [39].

Diapause in some species of flies is subject to both photoperiod and temperature; for example, the dominant diapause cues of *M. autumnalis* and *H. irritans* are probably both photoperiod and temperature [40]. The critical day length for *Aphidoletes aphidimyza* diapause induction is determined to be 12.7 h at 20°C. Diapause incidence is completely prevented at 30°C even though the photoperiod used is 11 L–13D. In addition, diapause induction is 100% under changing temperature conditions while maintaining the critical day length (12.7 L–11.3D), and diapause incidence is 100% in both field and greenhouse conditions under alternating temperatures of 20/16 or 25/16°C [41]. Winter diapause of *Delia antiqua* is completed under constant diapause-inducing conditions of 15°C and 12 L–12D, and the pupal period is shortened significantly [42]. The critical photoperiod for inducing diapause of *E. civilis* is between 11.8 and 11.9 h, while it can be induced into diapause at longer day lengths when the temperature is lower [43]. The incidence of *E. civilis* diapause is also influenced by the temperature and photoperiod. It is found that lower temperature prolongs the larval diapause period, and the diapause rate decreases under the high temperatures. All larvae enter diapause at 17°C, whereas the larvae become sluggish and stiff when temperature is lower than 17°C and death increases. The diapause rate is negatively correlated with light duration at 21°C. Diapause cannot be induced by short-day lengths at 25°C. Low temperature and short photoperiod are the most important prerequisites for inducing *E. civilis* diapause, among which temperature is the dominant factor, and photoperiod affects diapause induction only at certain temperatures.

7. Molecular mechanisms of diapause

7.1 Hormonal and metabolic regulation

Endogenous regulatory factors of insect diapause mainly focused on neuroendocrine systems, hormone signaling pathways, and energy metabolism pathways. Studies on *Drosophila* show that when insulin-like signals are disrupted, fruit flies stop reproducing, and their energy reserves increase [44]. Inhibiting FOXO and insulin-like polypeptide-1 (ILP-1) change many diapause phenotypes of *C. pipiens* such as increase stress resistance, fat accumulation, and delayed development [45]. Therefore, insulin signaling is thought to be an important candidate pathway to study differential regulation of diapause metabolism.

Juvenile hormone (JH), a sesquiterpenoid hormone produced by the corpus allatum (CA) of insects, is one of the most important hormones in insects and plays a key role in preventing larval metamorphosis, maintaining larval state, and regulating adult developmental and physiological process. It also plays a crucial role in the expression of vitellogenin, oocyte maturation, and development. The interaction between genes associated with juvenile hormone pathway is complex, which means genes directly or indirectly participate in the regulation of JH signaling pathway. Studies on the fly *Melinda pusilla* have reported that the application of exogenous juvenile hormone can temporarily terminate diapause [46]. Diapause in *C. pipiens* adult stage is characterized by a pause in reproduction, such as stagnation of the ovaries and male accessory glands, as well as reduced mating activity. The cessation of JH production delays ovarian development, and increases stress resistance, fat, and sugar storage [47]. When *C. pipiens* females are held in diapause conditions for 22 weeks, follicles gradually grow longer. When 21-day-old diapausing mosquitoes are moved to a long daylight of 16:8(L:H) at 26°C, juvenile hormone synthesis increases rapidly and peaks 5 days later, while follicles grow to a quiescent stage. Allatectomy of young diapausing females prevents follicle growth and blood feeding when diapause is terminated prematurely, demonstrating that the physiological events are associated with juvenile hormone biosynthesis [48].

During diapause, there are a lot of significant changes that occur in energy and metabolism due to the organism need to maintain life activities under extreme environmental conditions. Expression of the trehalase gene expression and enzyme activity of *D. antiqua* in summer- and winter diapause are lower at the initial phase but increase gradually and peak in the maintenance phase [49]. The contents of glycogen and trehalose in *E. civilis* during diapause increase significantly with an increase in the diapause induction period.

7.2 Diapause-associated changes in genes

With the development of high-throughput sequencing technology, the sequencing and annotation of *Drosophila melanogaster* genome were first published in 2000 [50]. However, not all Dipteran insects have been sequenced. To simplify the genome sequencing research strategy, transcriptome sequencing can be used to study insects with or without a reference genome available. Transcriptome sequencing technology is rapid, efficient, and low-cost, and currently has been widely used in genetic research, which can carry out transcriptome analysis of a tissue or organelle in a more comprehensive way [51]. Transcriptomes would not only facilitate a better understanding of how individual genes have evolved in biological processes, but would also help to reflect what gene expression changes more precisely [52]. The transcriptome refers to the collection of all gene transcription products in a cell or tissue under certain state or physiological condition, including messenger RNA, ribosomal RNA, transport RNA, and non-coding RNA. Transcriptomics, as an important means to study cell function, can be used to investigate the amount of active gene expression at the RNA level. In recent years, with the development of second- and third-generation molecular sequencing technology, the single-cell genomics is advancing rapidly. Transcriptome based on whole-genome analysis represents the average level of all cell signals, ignoring the variation between cells [53]. Compared with traditional sequencing technology, single-cell technologies have the advantages of detecting the differences in gene expression among individual cells. At present, single-cell sequencing has been used in *Drosophila* and mosquitoes. New cell types, mechanisms of development and aging, genes controlling neural regulation, and connection have been found in *Drosophila* by single-cell RNA sequencing analysis [54]. However, single-cell sequencing has not been done to study diapause of Dipteran insects.

The use of RNA-Seq to determine genes with distinct levels of expression between diapause and non-diapause has been confined to flies, and Kyoto Encyclopedia of Genes (KEGG) analysis is performed to identify the pathways that are significantly enriched in diapause. A high-throughput RNA-Seq analysis from non-diapause and summer diapause pupae of *D. antiqua* revealed variation of cuticular and cytoskeletal components [55]. Significantly differentially regulated transcripts are identified in summer diapause (SD) and non-diapause (ND)-sensitive larvae of *D. antiqua*. Several functional terms related to lipid, carbohydrate, and energy metabolism, environmental adaption, immune response, and aging are enriched during the most sensitive SD induction period. There is much more variation of circadian clock genes in the period of ND than SD-destined larvae, which indicates that it is a key driver of integrating environmental signals to summer diapause [56]. The diapause-associated traits identified in *S. mosellana* appear to be involved in rapid spread and outbreaks. Transcriptomic sequencing performed on diapause and non-diapause larvae shows that various genes-coding metabolic enzymes are crucial for diapause [57]. The molecular mechanisms of obligatory diapause induction in *B. minax* are investigated by using high-throughput RNA-Seq data from second-instar larva, third-instar larva, and pupa stages. The cluster co-expression patterns of the differentially expressed genes reveal that significantly differentially expressed genes in the pupal stage are predicted to be related to diapause induction. All differentially expressed genes are investigated by GO functional and KEGG pathway analysis, and the results show that genes involved in processes such as 20-hydroxyecdysone (20E) biosynthesis, cell cycle, and metabolic pathways are likely related to obligatory diapause induction in *B. minax* [58]. The diapause-associated genes in *E. civilis* are related to be involved in the pathway of signal transduction, endocrine system, and carbohydrate metabolism by KEGG pathway enrichment analysis [59].

Based on transcriptome sequencing, some candidate diapause-related genes have been further studied in Diptera. For example, heat shock proteins (HSPs) have been studied in *S. mosellana*, *Sarcophaga crassipalpis*, and *Rhagoletis mendax*. Hsp90 is downregulated in *S. crassipalpis* diapause pupae and returns to pre-diapause level after diapause termination. The expression of Hsp90 is increased by heat shock or cold shock during diapause. It is showed that Hsp90 is regulated differently in diapause and diapause pupae response to heat injury [60]. DaTrypsin is a serine protease gene and is the first upregulated gene during winter and summer diapause. It may be involved in host immune defense or maintain the developmental of diapause pupae [61].

7.3 Diapause-associated changes in proteins

Proteins are complex molecules that play a central role in biological processes. Proteomics is used to elucidate the expression and function of protein on the basis of genome research. Changes in protein expression during diapause can be explored by two-dimensional gel electrophoresis and mass spectrometry. Isobaric tag for relative and absolute quantitation (iTRAQ) can quantitatively analyze proteins from different sources in a single assay, and is used to study quantitative changes in the proteome by tandem mass spectrometry. Due to its high efficiency and sensitivity, iTRAQ has the potential to further advance the study of molecular mechanisms involved in diapause. Proteomic analysis of Diptera has also been reported. A proteomic approach was used to investigate the proteins extracted from larvae of *S. mosellana* at different developmental stages, which include pre-diapause, over-summering diapause, over-wintering diapause, and post-diapause. The results showed that two small Hsps play key roles in stress tolerance during diapause [62]. Proteins synthesized by pupal brains of the flesh fly *S. crassipalpis* were examined during diapause and non-diapause using pulse labeling and two-dimensional electrophoresis, and it was found that a cluster of about 15 brain proteins appears to be specific to diapausing pupae [63]. Ninety-five differently expressed proteins were identified in the diapause of *E. civilis* by using iTRAQ proteomics, and Hsps were found to be the key diapause-associated proteins. These proteins are mainly involved in material and energy metabolism [64]. Proteomic changes are studied in diapausing versus non-diapausing *D. antiqua* using two-dimensional differential gel electrophoresis, and it is found that some identified differential proteins may play an important role in physiological processes such as heat resistance, chromosome separation, and folic acid metabolism [65].

7.4 Diapause-associated changes in metabolite profiles

Metabolome refers to a collection of small molecular compounds that participate in the metabolism of an organism or cell with a relative molecular weight of less than 1000 DA in a specific physiological period. Metabolomics is a new discipline that simultaneously conducts qualitative and quantitative analysis of small molecule metabolite. It can be used to investigate how metabolites change with time when the organism is stimulated [66]. The cellular activities of living organisms are jointly undertaken by genes, proteins, and small molecule. The metabolic level can reflect the functional changes of macromolecules and amplify the small changes in gene expression. Various techniques are widely used to determine metabolic phenotypes, including liquid chromatography-mass spectrometry (LC-MS), gas chromatography-mass spectrometry (GC/MS), and nuclear magnetic resonance (NMR). LC-MS analysis does not require sample volatileness and thermal stability, and is suitable for compounds with high boiling points, strong polarity, and poor thermal stability. Most metabolites involved in life science have these qualities, so LC-MS has broad application. In recent years, metabolomics has been applied in the area of medicine and microbiology, but there are still relatively few studies on insect diapause. The metabolic profiles of diapause and non-diapause *B. minax* pupae show that proline, trehalose, N-acetylglutamate, and alanine significantly contribute to cold tolerance during diapause [67]. In the metabolomics analysis of diapause and non-diapause pupae of *E. civilis*, L-proline, L-phenylalanine, L-histidine, and L-tyrosine are significantly different, which provides a foundation for mechanistic follow-up studies in insect diapause [68].

8. Conclusion

Knowledge of diapause in Diptera is essential for the development of effective pest management strategies and to increase the shelf-life of parasitoids used in the biological control industry. This chapter summarizes the recent progress on diapause of Diptera. We do believe that further studies should be investigated in the diapause of Diptera. Current research studies suggest that histone modification, DNA methylation, RNA methylation, and small noncoding RNAs all may be involved in the regulation of diapause in Diptera. However, it remains unclear whether they regulate the hormonal and physiological changes associated with diapause of Diptera; research has primarily focused on physiological changes associated with pre-diapause with limited attention given to post-diapause. Studies showed that the indices of insects, such as oviposition quantity, oviposition duration, and life span, increased positively during post-diapause phase. The accumulation and consumption of energy storage substances in pre-diapause and diapause will affect the biological characteristics in post-diapause phase. Combined with biological characteristics in post-diapause, studies on development rate, feeding, individual size, diapause maintain environmental conditions, and nutritional supplements after the diapause are necessary. Existing studies on diapause in Diptera are mainly based on single omics, and studies using multi-omics are still vacant. Therefore, for a deeper understanding of the complex molecular landscape of diapause in Diptera, all the available omics data should be utilized in combination rather than treating them individually.

Acknowledgements

This study was supported by the Science and technology planning project of Inner Mongolia Autonomous Region (2021GG0057, 2020ZY0001, 2019BS03011).

Author details

Haibin Han[1*], Yanyan Li[2], Bo Zhang[3], Kejian Lin[1], Shujing Gao[1], Linbo Xu[1], Ning Wang[1], Wenchang Duan[2] and Wenyuan Niu[4]

1 Institute of Grassland Research, Chinese Academy of Agricultural Sciences, Hohhot, China

2 Inner Mongolia Agricultural University, Hohhot, China

3 Zhejiang University, Hangzhou, China

4 Garden Bureau of Chahar Right Front Banner, Ulanqab, China

*Address all correspondence to: hhb.25@163.com

References

[1] Denlinger DL. Regulation of diapause. Bulletin of the Entomological Society of America. 1979;**47**:93-122. DOI: 10.1146/annurev.ento.47.091201.145137

[2] Stålhandske S, Lehmann P, Pruisscher P, Leimar O. Effect of winter cold duration on spring phenology of the orange tip butterfly *Anthocharis cardamines*. Ecology and Evolution. 2015;**5**:5509-5520. DOI: 10.1002/ece3.1773

[3] Yokoyama T, Saito S, Shimoda M, Kobayashi M, Takasu Y, Sezutsu H, et al. Comparisons in temperature and photoperiodic-dependent diapause induction between domestic and wild mulberry silkworms. Scientific Reports. 2021;**11**(1):8052. DOI: 10.1038/s41598-021-87590-4

[4] Sim C, Denlinger DL. Insulin signaling and FOXO regulate the overwintering diapause of the mosquito *Culex pipiens*. Proceedings of the National Academy of Sciences. 2008;**105**(18):6777-6781. DOI: 10.1073/pnas.0802067105

[5] Hand SC, Denlinger DL, Podrabsky JE. Mechanisms of animal diapause: Recent developments from nematodes, crustaceans, insects, and fish. American Journal of Physiology-Regulatory Integrative and Comparative Physiology. 2016;**310**(11):R1193-R1211. DOI: 10.1152/ajpregu.00250.2015

[6] Lysyk TJ, Moon RD. Diapause recruitment and survival of overwintering *Haematobia irritans* (Diptera: Muscidae). Environmental Entomology, Annapolis: Entomological Society of America. 2001;**30**(6):1090-1097. DOI: 10.1603/0046-225X-30.6.1090

[7] Suman DS, Wang Y, Gaugler R. The insect growth regulator pyriproxyfen terminates egg diapause in the Asian tiger mosquito, *Aedes albopictus*. PLoS One. 2015;**10**(6):e0130499. DOI: 10.1371/journal.pone.0130499

[8] Takeda M. Photoperiodic induction, maintenance and termination of winter diapause in two geographic ecotypes of the rice stem maggot, *Chlorops oryzae* Matsumura (Diptera: Chloropidae). Applied Entomology and Zoology. 1996;**31**(3):379-388. DOI: 10.1007/BF01904716

[9] Amouroux P, Normand F, Delatte H. Diapause incidence and duration in the pest mango blossom gall midge, *Procontarinia mangiferae* (Felt), on Reunion Island. Bulletin of Entomological Research. 2014;**104**(5):661-670. DOI: 10.1017/S0007485314000480

[10] Zhai Y, Lin Q, Zhang J. Adult reproductive diapause in *Drosophila suzukii* females. Journal of Pest Science. 2016;**89**(3):679-688. DOI: 10.1007/s10340-016-0760-9

[11] Irwin JT, Bennett VA, Lee RE. Diapause development in frozen larvae of the goldenrod gall fly, *Eurosta solidaginis* Fitch (Diptera: Tephritidae). Journal of Comparative Physiology B-Biochemical Systemic and Environmental Physiology. 2001;**171**(3):181-188. DOI: 10.1007/s003600000154

[12] Tachibana SI, Numata H. Effects of temperature and photoperiod on the termination of larval diapause in *Lucilia sericata* (Diptera : Calliphoridae). Zoological Science. 2004;**21**(2):197-202. DOI: 10.2108/zsj.21.197

[13] Wise IL, Lamb RJ. Diapause and emergence of *Sitodiplosis mosellana* (Diptera : Cecidomyiidae) and its parasitoid *Macroglenes penetrans* (Hymenoptera: Pteromalidae). Canadian Entomologist. 2004;**136**(1):77-90. DOI: 10.4039/N03-032

[14] Košťál V. Eco-physiological phases of insect diapause. Journal of Insect Physiology. 2006;**52**(2):113-127. DOI: 10.1016/j.jinsphys.2005.09.008

[15] Wu JX, Li CQ, Li YP. Advances in research on diapause of the wheat blossom midges. Chinese Journal of Applied Entomology. 2004;**41**(6):499-503. DOI: 10.3969/j.issn.0452-8255.2004.06.001

[16] Hahn DA, Denlinger DL. Energetics of insect diapause. Annual Review of Entomology. 2011;**56**(1):103-121. DOI: 10.1146/annurev-ento-112408-085436

[17] Ren XY, Qi XY, An T. Research on the accumulation, transformation and regulation of nutrients in diapause insects. Journal of Applied Entomology. 2016;**53**(4):685-695. DOI: 10.7679/j.issn.2095-1353.2016.086

[18] Mitchell CJ, Briegel H. Inability of diapausing *Culex pipiens* (Diptera: Culicidae) to use blood for producing lipid reserves for overwinter survival. Journal of Medical Entomology. 1989;**26**(4):318-326. DOI: 10.1111/j.1365-2761.1989.tb00328.x

[19] Veerman A, Slagt ME, Alderlieste MFJ. Photoperiodic induction of diapause in an insect is vitamin A dependent. Experientia. 1985;**41**(9):1194-1195. DOI: 10.1007/BF01951725

[20] Paolucci S, Zande LV, Beukeboom LW. Adaptive latitudinal cline of photoperiodic diapause induction in the parasitoid *Nasonia vitripennis* in Europe. Journal of Evolutionary Biology. 2013;**26**(4):705-718. DOI: 10.1111/jeb.12113

[21] Powell CP, Bass MH. Some effects of photoperiod, temperature, and food on the induction of diapause in the Boll Weevil. Journal of Economic Entomology. 1968;**61**(3):624-625. DOI: 10.1093/jee/61.3.624

[22] Chen Z, Dong Y, Wang Y. Pupal diapause termination in *Bactrocera minax*: an insight on 20-hydroxyecdysone induced phenotypic and genotypic expressions. Scientific Reports. 2016;**6**:27440. DOI: 10.1038/srep27440

[23] Holbrook GL, Schal C. Maternal investment affects offspring phenotypic plasticity in *Aviviparous cockroach*. Proceedings of the National Academy of Sciences of the United States of America. 2004;**101**(15):5595-5597. DOI: 10.1073/pnas.0400209101

[24] Reznik SY, Voinovich ND. Maternal regulation of *Trichogramma embryophagum* diapause photoperiodic sensitivity of adult females. Biological Control. 2011;**57**(2):158-162. DOI: 10.1016/j.biocontrol.2011.02.005

[25] Melanie G, Hans VD, Bengt K. Reproductive plasticity, ovar-ian dynamics and maternal effects in response to temperature and flight in *Pararge aegeria*. Journal of Insect Physiology. 2010;**56**(9):1275-1283. DOI: 10.1016/j.jinsphys.2010.04.009

[26] Savalli UM, Czesak ME, Fox CW. Paternal investment in the seed beetle, *Callosobruchus maculatus* (Coleoptera: Bruchidae): Variation among populations. Annals of the Entomological Society of America. 2000;**93**(6):1173-1178. DOI: 10.1023/A:1005834807873

[27] Lacour G, Vernichon F, Cadilhac N. When mothers anticipate: Effects of the prediapause stage on embryo development time and of maternal photoperiod on eggs of a temperate and a tropical strains of *Aedes albopictus* (Diptera: Culicidae). Journal of Insect Physiology. 2014;**71**:87-96. DOI: 10.1016/j.jinsphys.2014.10.008

[28] McWatters HG, Saunders DS. Maternal temperature has different effects on the photoperiodic response and duration of larval diapause in blow fly (*Calliphora vicina*) strains collected at two latitudes. Physiological Entomology. 1998;**23**(4):369-375. DOI: 10.1046/j.1365-3032.1998.234101.x

[29] McWatters HG, Saunders DS. The influence of each parent and geographic origin on larval diapause in the blow fly, *Calliphora vicina*. Journal of Insect Physiology. 1996;**42**(8):721-726. DOI: 10.1016/0022-1910(96)00030-3

[30] Shin-Ichiro TACHIBANA, Hideharu. Maternal induction of larval diapause and its sensitive stage in the blow fly *Lucilia sericata* (Meigen) (Diptera: Calliphoridae). Entomological. Science. 2004;**7**(3):231-235. DOI: 10.1111/j.1479-8298.2004.00068.x

[31] Reynolds JA, Peyton JT, Denlinger DL. Changes in microRNA abundance may regulate diapause in the flesh fly, *Sarcophaga bullata*. Insect Biochemistry and Molecular Biology. 2017;**84**:1-14. DOI: 10.1016/j.ibmb.2017.03.002

[32] Emerson KJ, Bradshaw WE, Holzapfel CM. Complications of complexity: Integrating environmental, genetic and hormonal control of insect diapause. Trends in Genetics: TIG. 2009;**25**(5):217-225. DOI: 10.1016/j.tig.2009.03.009

[33] Armbruster PA. Photoperiodic diapause and the establishment of *Aedes albopictus* (Diptera: Culicidae) in North America. Journal of Medical Entomology. 2016;**53**(5):1013-1023. DOI: 10.1093/jme/tjw037

[34] Takeda M. Effects of photoperiod and temperature on larval development and summer diapause in two geographic ecotypes of the rice stem maggot, *Chlorops oryzae* Matsumura (Diptera: Chloropidae). Applied Entomology and Zoology. 1997;**32**(1):63-74. DOI: 10.1007/BF01991956

[35] Takeda M, Nagata T. Effects of temperature and winter diapause on survival and development in bivoltine and trivoltine ecotypes of the rice stem maggot, *Chlorops oryzae* Matsumura (Diptera: Chloropidae), reared on a winter host. Applied Entomology and Zoology. 1998;**33**(1):85-96. DOI: 10.1086/320066

[36] Goto SG, Numata H. Alteration of the pupal diapause program and regulation of larval duration by photoperiod in the flesh fly *Sarcophaga similis* Meade (Diptera: Sarcophagidae). Applied Entomology and Zoology. 2009;**44**(4):603-609. DOI: 10.1303/aez.2009.603

[37] Ichikawa A, Ikeda M, Goto SG. Cold storage of diapausing larvae and post-storage performance of adults in the blowfly *Lucilia sericata* (Diptera: Calliphoridae). Applied Entomology and Zoology. 2020;**55**(3):321-327. DOI: 0.1007/s13355-020-00685-8

[38] Zhou Z, Dong X, Su Q. Effects of pre-diapause temperature and body weight on the diapause intensity of the overwintering generation of *Bactrocera minax* (Diptera: Tephritidae). Journal of Insect Science. 2020;**20**(1):12. DOI: 10.1093/jisesa/ieaa005

[39] Numata H, Shiga S. Induction of adult diapause by photoperiod and temperature in *Protophormia terraenovae* (Diptera: Calliphoridae) in Japan. Environmental Entomology. 1995;**24**(6):1633-1636. DOI: 10.1093/ee/24.6.1633

[40] Fowler FE, Chirico J, Sandelin BA. Seasonality and diapause of *Musca autumnalis* (Diptera: Muscidae) at its southern limits in North America, with observations on *Haematobia irritans* (Diptera: Muscidae). Journal of Medical Entomology. 2015;**52**(6):1213-1224. DOI: 10.1093/jme/tjv123

[41] Masafumi Y, Eizi. Effect of photoperiod and temperature on the induction of diapause in a Japanese strain of *Aphidoletes aphidimyza* (Diptera: Cecidomyiidae). Applied Entomology and Zoology. 2012;**47**(1):17-26. DOI: 10.1007/s13355 011 0084-4

[42] Nomura M, Ishikawa Y. Biphasic effect of low temperature on completion of winter diapause in the onion maggot, *Delia antiqua*. Journal of Insect Physiology. 2000;**46**(4):373-377. DOI: 10.1016/S0022-1910(99)00120-1

[43] Hondelmann P, Poehling HM. Diapause and overwintering of the hoverfly *Episyrphus balteatus*. Entomologia Experimentalis Et Applicata. 2007;**124**(2):189-200. DOI: 10.1111/j.1570-7458.2007.00568.x

[44] Tatar M, Yin C. Slow aging during insect reproductive diapause: why butterflies, grasshoppers and flies are like worms. Experimental Gerontology. 2001;**36**(4):732-738. DOI: 10.1016/S0531-5565(00)00238-2

[45] Sim C, Denlinger DL. Insulin signaling and the regulation of insect diapause. Frontiers in Physiology. 2013;**4**:189. DOI: 10.3389/fphys.2013.00189

[46] Agui N, Mihara M, Kurahashi H. Effect of juvenile hormone analogue on ovarian development of the reproductive-diapausing parasitic fly *Melinda pusilla*. Medical Entomology and Zoology. 2016;**42**(4):311-317. DOI: 10.7601/mez.42.311

[47] Kang DS, Denlinger DL, Sim C. Suppression of allatotropin simulates reproductive diapause in the mosquito *Culex pipiens*. Journal of Insect Physiology. 2014;**64**:48-53. DOI: 10.1016/j.jinsphys.2014.03.005

[48] Janice R, Ming-Hsuan C, Roger M. Juvenile hormone biosynthesis in diapausing and nondiapausing *Culex pipiens* (Diptera: Culicidae). Journal of Medical Entomology. 1999;**36**(3):355-360. DOI: 10.1093/jmedent/36.3.355

[49] Guo Q, Hao YJ, Li Y. Gene cloning, characterization and expression and enzymatic activities related to trehalose metabolism during diapause of the onion maggot *Delia antiqua* (Diptera: Anthomyiidae). Gene. 2015;**565**(1):106-115. DOI: 10.1016/j.gene.2015.03.070

[50] Peng W, Feng MJ, Chen H. Progress on genome sequencing of Dipetran insects. Hereditas. 2020;**42**(11):1093-1109. DOI: 10.16288/j.yczz.20-130

[51] Zhang QL, Yuan ML. Progress in insect transcriptomics based on the next generation sequencing technique. Acta Entomologica Sinica. 2013;**56**(12):1489-1508. DOI: 10.16380/j.kcxb.2013.12.005

[52] Stark R, Grzelak M, Hadfield J. RNA sequencing: The teenage years. Nature Reviews Genetics. 2019;**20**(11):631-656. DOI: 10.1038/s41576-019-0150-2

[53] Zeb Q, Wang C, Shafiq S. An Overview of Single-Cell Isolation Techniques. USA: Elsevier, Academic

Press; 2019. pp. 101-135. DOI: 10.1016/B978-0-12-814919-5.00006-3

[54] McLaughlin CN, Brbic M, Xie Q. Single-cell transcroptomes of developing and adult olfactory receptor neurons in *Drosophila*. eLife. 2021;**10**:e63856. DOI: 10.1101/2020.10.08.332130

[55] Hao YJ, Zhang YJ, Si FL. Insight into the possible mechanism of the summer diapause of *Delia antiqua* (Diptera: Anthomyiidae) through digital gene expression analysis. Insect Sci. 2016;**23**(3):438-451. DOI: 10.1111/1744-7917.12323

[56] Ren S, Hao YJ, Chen B. Global Transcriptome sequencing reveals molecular profiles of summer diapause induction stage of onion maggot, *Delia antiqua* (Diptera: Anthomyiidae). G3-Genes Genomes. Genetics. 2018;**8**(1):207-217. DOI: 10.1534/g3.117.300393

[57] Gong ZJ, Wu YQ, Miao J. Global Transcriptome analysis of orange wheat blossom midge, *Sitodiplosis mosellana* (Gehin) (Diptera: Cecidomyiidae) to identify candidate transcripts regulating diapause. PLoS One. 2013;**8**(8):e71564. DOI: 10.1371/journal.pone.0071564

[58] Zhou Z, Dong X, Li C. Comparative transcriptome analysis of different developmental stage of *Bactrocera minax* (Diptera: Tephritidae): Implication of the molecular basis of its obligatory diapause induction. Comparative Biochemistry and Physiology D-Genomics and Proteomics. 2021;**38**:100818. DOI: 10.1016/j.cbd.2021.100818

[59] Zhang B, Han HB, Xu LB. Transcriptomic analysis of diapause-associated genes in *Exorista civilis* Rondani (Diptera: Tachinidae). Archives of Insect Biochemistry and Physiology. 2021;**107**(2):e21789. DOI: 10.1002/arch.21789

[60] Rinehart JP, Denlinger DL. Heat-shock protein 90 is down-regulated during pupal diapause in the flesh fly, *Sarcophaga crassipalpis*, but remains responsive to thermal stress. Insect Molecular Biology. 2000;**9**(6):641-645. DOI: 10.1046/j.1365-2583.2000.00230.x

[61] Chen B, Kayukawa T, Jiang HB. DaTrypsin, a novel clip-domain serine proteinase gene up-regulated during winter and summer diapauses of the onion maggot, *Delia antiqua*. Gene;**347**(1):115-123. DOI: 10.1016/j.gene.2004.12.026

[62] Zhao J, Huang Q, Zhang G. Characterization of two small heat shock protein genes (Hsp17.4 and Hs20.3) from *Sitodiplosis mosellana*, and their expression regulation during diapause. Insects. 2021;**12**(2):119. DOI: 10.3390/insects12020119

[63] Joplin K, Yocum G, Denlinger D. Diapause specific proteins expressed by the brain during the pupal diapause of the flesh fly, *Sarcophaga crassipalpis*. Journal of Insect Physiology. 1990;**36**(10):775, 781-779, 783. DOI: 10.1016/0022-1910(90)90051-G

[64] Han HB, Jiang S, Lin KJ. Proteomic analysis of diapause proteins in the dominant parasitoid tachinid fly *Exorista civilis*. Journal of Plant Protection. 2021;**48**(6):1359-1369. DOI: 10.13802/j.cnki.zwbhxb.2021.2021892

[65] Jiang T, Zhu YL, Peng YC. Universal and differential transcriptional regulatory pathways involved in the preparation of summer and winter diapauses in *Pieris melete*. Bulletin of Entomological Research. 2021;**111**(3):371-378. DOI: 10.1017/S0007485321000018

[66] Dunn WB, Broadhurst D, Begley P. Procedures for large-scale metabolic profiling of serum and plasma using gas chromatography and liquid chromatography coupled to mass spectrometry. Nature Protocols. 2011;**6**(7):1060-1083. DOI: 10.1038/nprot.2011.335

[67] Wang J, Fan H, Xiong KC. Transcriptomic and metabolomic profiles of Chinese citrus fly, *Bactrocera minax* (Diptera: Tephritidae), along with pupal development provide insight into diapause program. PLoS One. 2017;**12**(7):e0181033. DOI: 10.1371/journal.pone.0181033

[68] Zhang B, Han HB, Xu LB. Metabolomic analysis of diapause-associated metabolites in tachinid parasite *Exorista civilis* based on liguid chromatograph mass spectrometry. Journal of Plant Protection. 2022;**49**(2):705-706. DOI: 10.13802/j.cnki.zwbhxb.2021.2020228

Section 2

Conventional and Advanced Strategies for Mosquito Management

Chapter 2

Innovative Methods of Mosquito Management

Zeeshan Javed, Saira Mansha, Usama Saleem, Asad Mangat, Bilal Rasool, Muhammad Imran, Amna Batool, Mashal Shahzadi, Tehreem Raza, Danish Riaz and Muhammad Asrar

Abstract

As the global human and animal population increases, deadly pathogens and parasites may be transmitted by arthropods. There are a number of vectors that pose a threat to human health due to their role in transmitting dangerous pathogens, including mosquitoes (Diptera: Culicidae). The most important drawback of these products is the incidence of insecticide resistance, which has increased rapidly in recent years. New approaches and vector-control tools targeting aquatic stages and adults are urgently needed. The three main mosquito genera, *Anopheles, Aedes, and Culex,* transmit the causative agents of numerous important diseases to humans as well as animals. A technique that involves the use of genetically modified (GM) mosquitoes for the purpose of vector control is another potential option. Other best ways to control the mosquito are by chemical, biological and genetic means.

Keywords: populations, pandemic, pathogens, approaches, causative

1. Introduction

There are several causes that contribute to the close relationship between fauna and flora in the globe today, including population expansion and the development of transportation networks. These variables break down biogeographic boundaries and lead to the initial occurrence of species in new environments [1]. Damage caused by these species in the Americas is estimated to be in the neighborhood of $120 billion each year [2]. Arthropods may spread deadly infections and parasites [3]. Epidemics and pandemics present a risk to the world's growing human and animal populations [4]. It is important to note that mosquitoes (Diptera: Culicidae) are considered the most harmful vector because of their involvement in the transfer of deadly infections [5]. As a result of commerce and travel, essential mosquito species are being introduced to new environments [6, 7]. Commercially accessible chemical compounds, employed for mosquito management, are often toxic to human beings imparting major skin and nervous system issues, such as rashes, swelling, or eye irritation [8]. Since pesticide resistance has been on the rise in recent years [9], treating mosquito breeding is exceedingly difficult or impossible, as these products pose a serious problem. Thus, the need for new vector control techniques and instruments that target aquatic stages and adults is critical [10]. Currently available information about mosquito-borne illnesses, as well as the most recent statistics on their reappearance, are discussed in this chapter, along with the strengths and weaknesses of the management methods currently in practice. For biological control of mosquito-borne illnesses, new inventive alternatives are recognized but seldom used, others that are not employed at all, and rest are in the test or design phase.

Mosquitoes of the three major genera *Anopheles*, *Aedes*, and *Culex* spread the pathogens that cause a wide range of serious illnesses in both people and animals [11]. There has been an increase in the number of mosquito-borne diseases that have a significant influence on human health.

As the most common parasite illness of humans, malaria, transmitted by *Anopheles* sp. is prevalent in more than 80 countries, with the majority of cases and fatalities occurring in sub-Saharan Africa, where more than 85 percent of cases and 90 percent of deaths occur, largely in children under 5 years old (**Figure 1**). There have been 228 million cases of malaria at the globe level, with 93% of them occurring in Africa alone, and recent epidemics have decimated several places [12].

Many mosquito species transmit *Wuchereria bancrofti* and *Brugia spp*., leading to lymphatic filariasis which produce diverse clinical symptoms (25 million men with hydrocele and over 15 million individuals with lymphoedema), and at least 36 million people continue to have these long-lasting illness manifestations. It is apparent, however, that eradicating lymphatic filariasis is not achievable without managing the vectors of this disease.

Likewise, *Aedes aegypti* and *Aedes albopictus* (**Figure 2**) transmit dengue, yellow fever, Chikungunya Zika, etc. in human beings. The dengue virus (DENV), containing four unique serotypes, belong to the Family Flaviviridae [13]. With 3.6 billion people living in places at risk of transmission and hundreds of millions of cases of dengue fever recorded each year [14], it is now the most common arthropod-borne viral illness affecting humans [15].

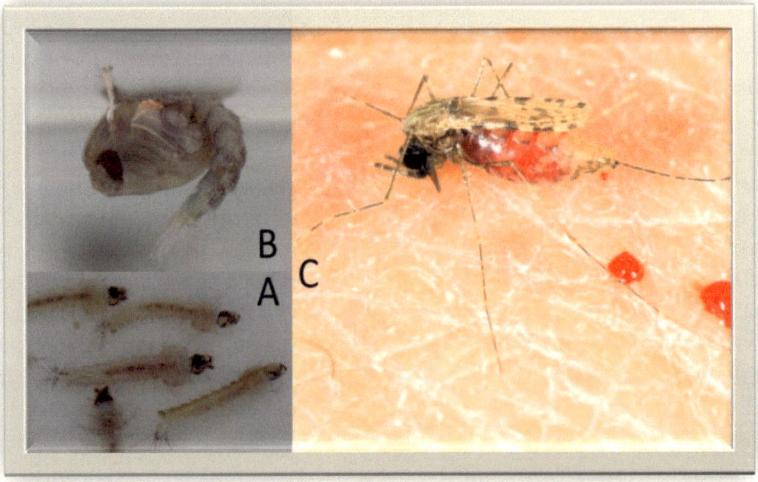

Figure 1.
Anopheles gambiae: (A) larvae, (B) pupa, and (C) adult.

DOI: http://dx.doi.org/10.5772/intechopen.107364

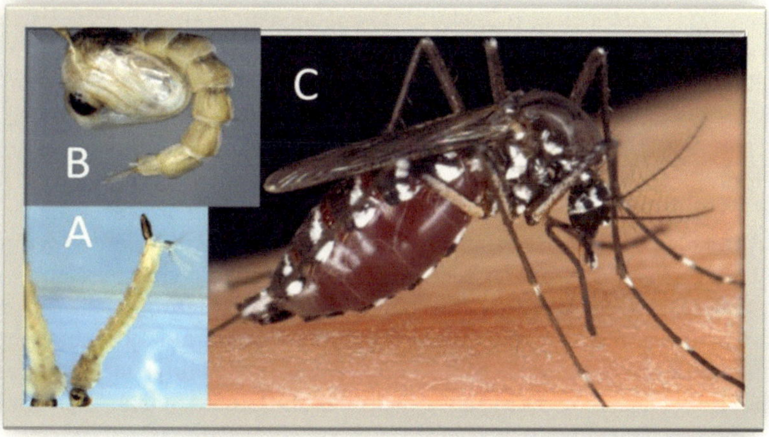

Figure 2.
Aedes albopictus: (A) larvae, (B) pupa, and (C) adult.

Zika virus (ZIKV) also belonging to Flaviviridae causes frequent disease outbreaks in numerous Latin American and Pacific nations (**Figure 3**) [16]. *A. albopictus* is the major and secondary vectors of ZIKV epidemics [17]. In addition, a number of other species are also implicated in the fast spread of this virus. It is now regarded as one of the most significant public health threats [18, 19].

Antalgic stance gait and acute articular pain are hallmarks of chikungunya fever (CHIF) caused by the Chikungunya virus (CHIKV) (**Figure 3**). The 1.4 percent to 90 percent of infected individuals often progress to the chronic stage (52 percent in the American continent). Several nations have lately had outbreaks of the disease [20].

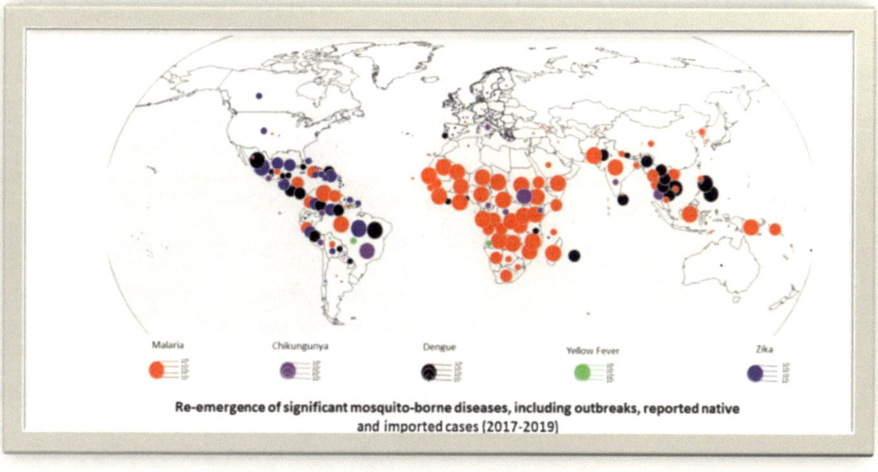

Figure 3.
Cartography of significant resurgences of mosquito-borne diseases worldwide (until September 2019). It presents outbreaks and cases of malaria, dengue fever, yellow fever, chikungunya fever, and Zika fever between 2017 and 2019. The figure clearly illustrates their resurgence in most tropical countries. There are several northern countries where the competent vector has become established, which may allow local transmission.

Apart from these diseases, the hemorrhagic and possibly deadly RNA virus, Flaviviridae [21], causes more diseases and generates more epidemics in various countries, particularly in unprotected populations [21]. A new epidemic usually appears every 7–10 years or so [22]. More than half the YFV (Yellow Fever Virus)-endemic nations in the world are located in Africa, according to the World Health Organization (WHO). Numerous outbreaks have been reported, with a death rate of up to 33.6% [23]. Roughly 70 to 90 million doses of vaccines are manufactured each year across the globe, making it the most cost-effective and safe method of preventing YF.

According to the World Health Organization, around 67,000 people per year are infected with *Culex*-borne Japanese encephalitis, 20 percent to 30 percent of whom die, and 30 to 50 percent of those who survive have major neurological consequences. As previously mentioned, new strains that are genetically similar to those implicated in prior epidemics have been discovered [24]. Louis encephalitis virus was the primary arbovirus cause of epidemic encephalitis in the United States [25]. Numerous people have been struck down by it in the last few years [26].

Similar to humans, horses are the domesticated animal most frequently infected by the West Nile virus transmitted by *Culex* (**Figure 4**). Eighty percent of cases are asymptomatic, neurological signs are the most frequently reported symptom, and 90 percent of those infected with the virus go on to develop clinical symptoms. The mortality rate for infected horses may reach 30 percent [24]. In spite of this, recent cases of human epidemics have been documented [25].

It is common to find different harmful blood-borne bacteria in mosquitoes [27]. It is not yet known whether the existence of these bacteria in mosquitoes may be attributed to their infrequent consumption of blood meals or environmental acquisition, or if they can grow and ultimately transmit during blood meals. Various pathogenic alpha-proteobacteria, such as *Anaplasma species*, *Ehrlich species*, *Candidatus neoehrlich species*, *Bartonella species*, and *Rickettsia species*, have been discovered in adult mosquitoes (xeno-monitoring studies) [28]. More intriguingly, laboratory tests have shown

Figure 4.
Culex pipiens: (A) larvae, (B) pupa, and (C) adult.

that the agent of febrile rickettsiosis, *Rickettsia felis*, may be transmitted by *Anopheles* mosquitoes [29]. *Francisella tularensis* is the first known mosquito-borne bacteria since it is also transmitted by mosquitoes (*Aedes*), the primary vector in Sweden and Finland [30].

Inadequate vector-control efforts, limited access to quality healthcare, rapid and unplanned urbanization of tropical regions coupled with unsanitary conditions, a deterioration of public health infrastructures, and a number of complex factors, including population growth, globalisation of the economy, international travel (recreational, business, and military), may explain the spread of these diseases in the region. All of these factors are connected to climate change [31]. The misuse of insecticides and the development of resistance, however, continue to be the primary contributing causes.

2. Mosquito control

Controlling the mosquito vectors that spread the major diseases is an important part of global initiatives to get rid of and control diseases. If these initiatives are successful, they can lead to a huge drop in the number of diseases around the world. However, various problems are associated with the vector control measures, which are discussed below.

The prevalence of pesticide resistance among vectors is a major obstacle that prevents the proper implementation of vector control measures [32]. Adult vector mosquitoes may be particularly difficult to manage if insecticides are no longer effective, leading to serious implications for human health. Another issue is the wide range of mosquito susceptibilities to various pesticides [33]. Malaria has been reduced by the use of ITNs (Insecticide-treated bed nets) and IRS (Indoor Residual Sprays) but insecticide resistance and the inability to maintain these treatments might have the inverse effect. Some ITNs and IRS approaches have shown potential in malaria control, but are restricted in dengue control because of the ecology of *Anopheles* and *Aedes* mosquitoes. The enormous variety of vectors and the changing behavior of mosquitoes are a few more obstacles to overcome. In addition, rising urbanization and climate change [24] might have an unanticipated and significant effect on the distribution of vectors, creating new challenges for vector management strategies [34–36].

In addition to this, other challenges in the implementations of vector control initiatives include issues arising in prevention strategies such as a limited amount of funds or fair distribution of funds for vector control. These are just some of the challenges that are associated with the implementation of vector control programs. It is also possible for the vector control interventions to be weakened if there is a lack of effective monitoring methods relevant to pesticide resistance and the behavior of vectors. In addition, the absence of cooperation between governmental and non-governmental groups may have an impact on the vector control operations that are carried out. Migration of both people and products presents difficulties in terms of both disease emergence and vector control [37].

Although the approaches to eradicate mosquito-breeding places are very successful, they are not practicable in locations with intermittent water supply and if these methods are not applied at the grass-root level, the efficiency of these strategies is impaired [38]. For example, kerosene oil and chemical larvicides are excellent in killing larvae, but this method has a big drawback of being harmful to the environment. Similarly, application of the soil bacteria *Bacillus thuringiensis israelensis* (Bti) though effective, but works only on the larval stages and therefore using Bti is not a viable option for controlling larval populations. Further, Bti has been proven to have an influence on food chains and led to other environmental consequences in recent research [39, 40]. Likewise, the introduction of *Tilapia* and mosquito fish into the ecosystems without the use of a controlled ecosystem might potentially result in harmful environmental impacts.

The harmful effects of these chemical larvicides necessitate that the World Health Organization (WHO) understands and assesses the development of new vector control products before they are used in the field. Consequently, a variety of various vector control tactics and research are addressed below, such as the release of sterile insects by irradiation, the use of *Wolbachia* and gene-drive technologies, etc. Public acceptance and regulatory approval need a detailed risk assessment and significant stakeholder input.

While using genetically modified organisms, these kinds of hurdles are obviously far greater than they are for purely biological control methods like use of *Wolbachia*. Second, the implementation of these treatments will put enormous selection pressures, which might lead to the development of resistance in either the target pathogen (in the case of *Wolbachia* or vector competence gene drive constructs) or the vector (for population suppression genetic constructs and possibly *Wolbachia*). In addition, the long-term phenotypic stability of *Wolbachia* in *Ae. aegypti* is yet unclear at this time [41].

A technique that involves the use of genetically modified (GM) mosquitoes for the purpose of vector control is another potential option. This method has various benefits, such as the fact that it is non-toxic and prevents the use of chemical pesticides. However, the use of genetically modified mosquitoes raises a number of ethical difficulties. In addition to this, it is necessary to take into mind the possible influence that these creatures may have on the surrounding ecosystem [42]. In addition, the process of creating genetically modified mosquitoes is highly pricey and may not be feasible for economically disadvantaged endemic nations. Thus, the World Health Organization (WHO) advises and urges more field tests and risk assessments to determine how effective this method is in preventing the spread of disease [43].

Recently, the environmentally friendly production of nanoparticles has developed as an approach for vector control that is both easy and efficient in its use. The large-scale synthesis and its potential effects on the environment are, however, subject to a number of restrictions. In addition, there is a significant divide between the implications that may be theoretically drawn from this technology and those that can be drawn from its actual use. In addition, there is a very limited amount of data available about the effect that these nanoparticles have on other aquatic creatures [44]. A good number of these nanoparticles have been subjected to acute toxicity testing on non-target creatures or on other aquatic organisms that live in the same biological niche as vector mosquitoes.

The presence of high-transmission hotspots and heterogeneity both make the task of mosquito management more difficult; yet, it is possible that the task of control in low-transmission regions will be simpler than was originally anticipated based on geographically inaccurate transmission intensity forecasts [45].

Making the most of limited resources (especially in low-income regions) in order to achieve the greatest possible improvement in public health is going to be another one of the most challenging tasks. Extrapolation of clinical trial data to forecast population effect of each intervention in a wider variety of contexts and in conjunction

DOI: http://dx.doi.org/10.5772/intechopen.107364

with other control methods would require rigorous epidemiological research and mathematical modelling to ensure such optimal deployment. In addition, stringent monitoring and evaluation are necessary in order to determine whether therapies are beneficial in actual practice [36]. Meanwhile, the political commitment and implementation of collaborative vector control techniques are the keys to achieving the aim of vector control and, as a result, lowering the risk of disease transmission and making a contribution to the elimination of disease.

3. Alternative strategies for mosquito control

As many as 4 billion people are at danger of dengue virus transmission alone, despite our best attempts to manage vector-borne disease outbreaks using present intervention strategies. Because of this, new tactics for mosquito vector management must be devised in light of the current situation. With the fast rise in pesticide resistance and its harmful effects on non-target species, new methods for mosquito control are needed.

3.1 Repellents of the physical environment

To prevent interaction between people and the vector, the repellent compounds function in the vapor phase to make the area undesirable for the insect. Using this strategy, mosquitoes will be diverted to non-human hosts, reducing the damaging effects of pesticides on people and other non-target organisms, as a result. Instead of killing the insect, the repellents aim to keep it from biting humans [46]. New active chemical components can be used to change the typical behavior of the vectors and boost this method's effectiveness. Currently, there is no indication that this practice has any effect on the population. However, many hurdles must be overcome before spatial repellents may be used as a tool in vector management, as they cost too much money. Also, these repellents need the usage of power, which makes them unsuitable for places with a high transmission rate in less developed areas. If these deterrents are to be easily included into vector control programs, their cost must be comparable to that of IRS or LLINs (Long lasting insecticidal nets) [47].

Allethrin (Therma CELL) and metofluthrin (OFF! clip-ons or lamps) emanators have been tested in several early field trials for their efficiency as spatial repellents and have been found to provide more than 70% protection [48]. In push-pull systems, the application of these deterrents eventually aids in the mosquito's push towards the baited traps. Some of these trials involved the use of various repellents like as PMD, catnip oil, and delta-undecalactone.

3.2 Use of repellents derived from plants

Plant-based "natural" smelling insect repellents are now extensively used over the world since they are considered safe and effective. Many plant volatiles are effective insect deterrents and repellents due to their high vapor toxicity. Compounds present in most plants protect them from phytophagous (plant-eating) insects from being eaten. Among the compounds utilized are repellents, growth regulators, poisons, and feeding deterrents [49]. Alkaloids, terpenoids, proteinase inhibitors and phenolic compounds are among the greatest examples of secondary metabolites in plants which defend the plants. Currently, the volatile components produced by herbivory are well known for their capacity to repel mosquitoes and other biting pests. Insects detect volatile scents with the help of sensory neurons (ORNs), which are typically located on the antennae and maxillary palps of insects, and are equipped with odorant receptor (OR) proteins [50].

Many people have known for millennia that lemon eucalyptus has insect repellent properties. Essential oils containing 85 percent citronellal are far more effective than water in keeping mosquitoes away for long periods. The low vapor pressure of one of its ingredients, para-menthane-3, 8-diol, on the other hand, provides great protection against a wide range of insect vectors over an extended length of time. Eucalyptus extracts have recently been given fresh life by advances in nanotechnology [51]. As a repellent, citronella and vanillin are found in quantities of 5–10 percent in lemongrass extract and essential oil extract (5 percent). Using citronella oil in a nano-emulsion, stable droplets may be formed that help retain the oil and postpone its release. The efficacy of neem-based treatments has also been proven in various field studies in India [52].

3.3 Traps

Traps can be used to catch adult mosquitoes. The carbon dioxide that is released on conversion of propane into water can work as an attractant. Warm water vapors with carbon dioxide attract mosquitoes. As a result, mosquitoes may be lured up to 30 meters away from the trap using the pesticide octenol, or 1-octen-3-ol. This attractant mostly attracts zoophagous mosquitoes. In certain traps, a dim light is employed to attract the mosquitoes which are then into a collection chamber or bag using a fan. It is common in some mosquito traps, such as CDC light trap. This trap will catch many other flying insects, such as flies and bumblebees.

UV/visible light mosquito traps draw not only mosquitoes, but also helpful in catching pollinating insects, resulting in unintended injury. A larvicide pharmaceutical package is discharged to avoid the death of undesired insects; nevertheless, attracted insects may provide falsely optimistic image processing results. It is also possible to reduce power consumption by eliminating active traps that need actuators [42]. In order for a trap to be effective, it has to be set up, maintained, and operated correctly. Their effectiveness may be affected by the wind. An awkward location for a mosquito trap may lead to more attacks. To counter this problem, one might try setting up traps around their property [53].

3.4 Sterile insect technique (SIT)

The SIT is a pest control method that uses mass-reared sterile males to control an insect population in a certain region. When wild sterile males mate with wild females, no offspring are produced [54]. The target wild insect population is reduced over time by introducing sterile males in a systematic and repeated method. Various nations, including Brazil, Cuba, Italy, Mauritius, Mexico, and Germany, have tested the SIT on a small scale in partnership with the Food and Agriculture Organization (FAO) of the United Nations. The IAEA, TDR (Special Programme for Research and Training in Tropical Diseases), and WHO collaboration are planning larger-scale pilot releases as a part of research and technical cooperation activities, as well as test releases in connection with epidemiological investigations. There are two types of mosquitoes: those that bite and transfer diseases, and those that do not bite and do not constitute a threat to humans. Because sterile mosquitoes are unable to breed, their numbers will not rise in the wild. Normally, sterile mosquitoes are dispersed by land, but the IAEA, in partnership with the FAO and others, has achieved promising results in Brazil utilizing a drone release approach [51].

Tsetse flies, melon fruit fly, pink bollworm and the New World screwworm have all been eradicated using this method. Sterilization and genetic sexing, the development of superior strains for mass manufacturing and release, as well as the identification of molecular markers for detecting the released sterile insects in the field, can all help to increase the effectiveness of this procedure. To evaluate the SIT programme, it is critical to distinguish between sterile and wild insects that have been released [55]. A luminous transformation marker incorporated into a transgenic bug might make releasing insects easier to identify. *Anopheles gambiae* and *Aedes aegypti* mosquito species have evolved fluorescent sperm marking systems [51].

4. Management control of mosquitoes

Malaria, Mayaro fever, dengue, Chikungunya, yellow fever, filariasis, Zika, are just a few of the diseases that mosquitoes potentially spread. With the use of insecticides, larvicidal agents and bed nets, along with the use of medications as chemoprevention and treatment of the sick, these vectors can be control effectively.

Figure 5 represents the chemical, genetic and biological control techniques for mosquitoes which are discussed below.

Figure 5.
Chemical, biological and genetic control of mosquitoes.

4.1 Chemical control

Mosquitoes are responsible for the death of millions of people each year from diseases transmitted by vectors. There is currently no vaccine available for viral illnesses and malaria transmitted by mosquitoes. As a result, mosquito and vector species management is critical if epidemics like malaria and dengue fever are to be kept in check.

Control strategies based on chemical, biological and physical elements have all been implemented to prevent the spread of mosquito-borne disease as the most typical and conventional method of regulating mosquito populations. Among these, chemical control is the most productive, but it damages the environment and threatens non-target individuals as well. Despite their well-known negative effects, chemical insecticides combined with personal protection techniques are currently the most frequently applied strategy for mosquito control [56].

Chemicals with mosquitocidal qualities are known as insecticides. These chemicals include organochlorines, organophosphorus, carbamates, pyrethroids, pyrroles, and phenyl pyrazole. These substances are employed in sprays for public health purposes. The application of chemical pesticides as principal agents in excessive amounts without limit, without interference, without discriminatory treatment, and on a continuous basis result in warranted toxic or lethal effects on non-target organisms, resistance in mosquitos and most importantly the potential for toxic effects on environment and adverse effects on health, posing a great threat to life and the environment [57].

Pyrethroid pesticides are neurotoxic because they interfere with voltage-gated sensitive sodium channels (VSSC). Pyrethroids have a greater effect on insect sodium channels than on vertebrate sodium channels [58]. Pyrethroids are mixed with water or oil and applied as an ultra-low volume spray to kill flying adult mosquitoes by skilled mosquito control services. Toxic effects of pyrethroids are attributed to their ability to delay the activation of the voltage-sensitive sodium channel, which leads to immobility and eventually death of the insect, an effect known as "knockdown" [59]. Pesticides incorporating pyrethroids are most often used in the various countries to suppress dengue virus vectors *Aedes albopictus* (Skuse) and *Aedes aegypti* [60].

In the programmes, actively combating malaria and reducing the lifespan of gravid female mosquitoes, DDT (dichlorodiphenyltrichloroethane) may have been the most frequently employed man-made organic pesticide during the twentieth century. Water-based larvicides are used to control the number of larvae in the environment. Adult mosquito populations can be managed with adulticides and synergists, which disguise and spray adult mosquitoes. There are numerous insect development regulators, including pyriproxyfen, diflubenzuron, and methoprene, which can be used as larvicides and adulticides, along with ovicidal attributes, in mosquito control techniques worldwide [61].

In chemical control, the most obvious issues are growing pesticide resistance, human health hazard and the pollutants that has a detrimental effect on wildlife and the environment. Propoxur, permethrin, malathion, deltamethrin and lambda-cyhalothrin have been linked to behavioral changes in *Culex quinquefasciatus* and *A. aegypti* after exposure to sub-lethal concentrations of these organic chemicals, which belong to the three primary chemical classes; pyrethroids, carbamates, and organophosphates. Females of *A. aegypti* and *C. quinquefasciatus* species avoided feeding during the WHO tunnel experiments that used treated bed nets after a single sub-lethal pesticide exposure. The ability of mosquitos to change and resist towards insecticides has been shown to have a major influence on the effectiveness of these interventions [62]. As a result, there is an urgent need for further viable supply to synthetic insecticides over the world.

4.2 Biological control

Biological control has evolved from a specialized technique to a broader one over time. As a consequence, the number and variety of biocontrol agents used to treat pests and mosquito transmission has increased tremendously in recent decades [63].

Numerous studies have shown that a number of environmentally safe natural substances have insecticidal properties, including bioactive peptides, essential oils [64], nanomaterials [65] and polyphenolic extracts. Both natural enemies of the target mosquitoes and biotoxins are used in biological control tactics. Invertebrate predators, nematodes (such as *Romanomermis culicivorax*), larvivorous fish (such as *Gambusia affinis*), some fungi (like *Lagenidium giganteum*), and protozoa are included in this regard [66].

Shrimps of the species *Macrobrachium brasiliense, Macrobrachium amazonicum, M. pantanalense*, and *Macrobrachium jelskii* can be used to suppress *A. aegypti*, *C. quinquefasciatus* and *Anopheles darlingi* larvae [67].

As a result of the management of vectors, biocontrol approaches have also already contributed in reducing the mosquito number. As a biocontrol agent, bacteria such as *Bacillus thuringiensis*, *Bacillus sphaericus* and *Streptomyces avermitilis* have been shown to be safe for the environment and a feasible approach for reducing mosquitoes. One of the insecticidal protein families found in *B. thuringiensis* subsp. *israelensis*, the Cyt (cytolysins), as well as the Cry (crystal delta)–(endotoxins), have been proven to be efficient in mosquito control [68].

In specific African regions, some fungi are capable of attacking *A. aegypti*, which can be exploited as cost effective and ecofriendly for controlling flavivirus pandemics in North as well as South America [69]. *Microsporidia coelomomyces* is virulent and is known as the most varied parasitic fungus groups against mosquitos [70].

The use of larvivorous fish as a biological and self-reproducing adversary of insects through the process of predation is not only an extremely cost-effective approach of regulating mosquito populations, but it also has a mosquito control effect that is maintained over the long term. One of the fishes that is being employed the most often as a biocontrol agent is *Gambusia* [71]. It is clear that mosquito fish biocontrol has many benefits over traditional pesticide mosquito control mechanisms. Furthermore, employing exotic mosquito fish presents a problem because they might affect native species and local ecosystems. As a result, extreme care must be exercised while using them [72].

Toxorhynchites splendens is a mosquito genus well known for its larvae's capacity to feed on mosquito species such as *C. quinquefasciatus, A. darlingi* and *A. aegypti* as well as other aquatic animals living in naturally or artificially created habitats. *Toxorhynchites* mosquito is thought to be a potential biological control tool for mosquitos residing in a variety of habitats [73].

Insects that live in water, such as the *Lethocerus americanus*, have the potential to consume mosquitoes throughout various stages of their development [74]. Frogs, tadpoles, and toads can feed on the larvae of *A. darlingi*, *C. quinquefasciatus*, and *A. aegypti*.

Rhabditidae and Heterorhabditidae worms are known to either directly or indirectly cause death in their hosts. Nematodes can infect their hosts in one of four ways:

by entering the body while the host is just being fed by a mosquito; by penetrating the cuticle; by entering through the anus or spiracle; or by entering during mosquito feeding [75].

The plant extract-based larvicidal pesticides are a promising class because of their low toxicity, low environmental impact, and lack of harm to non-targeted species [76]. Mosquitocidal activity of different plants is combined with the microbicide capabilities of silver nanoparticles (Ag–NPs), leading to an improved nanoscale (1–100 nm) effectiveness due to the increased A/v of nanoparticles. Because of these attributes, Ag–NPs are remarkably efficient against vector larvae even at extremely low concentrations [77].

4.3 Genetic control

The number of disease-carrying mosquitoes can also be controlled by genetic methods that can target both the adults and the larvae [78]. For gene functional analysis and pest control, genome editing is essential. Orco (odorant receptor co-receptor) is a critical modulator of several olfactory-driven behaviors throughout the *Anopheles sinensis* life cycle, emphasizing the role of Orco as a potential molecular target for malaria control [79].

As an alternative to sterilizing males for the purpose of insect population control, genetic engineering is often used to introduce a gene into mosquito vectors that causes them to die off on their own [80]. It was discovered that OX513A males in Malaysia had similar lifespans [81] and spreading capacities, however the most recent production of OX513A males in Brazil ended in a large drop of the target wild population.

LA513A, an *A. aegypti* strain modified to possess a non-sex specific, dominant, complicated late acting lethal genetic process that leads to death during the pupal stage rather than larval maturation, avoids density dependent effects on larval stage in natural populations. Despite the fact that without tetracycline, the vast majority (95–97 percent) of LA513A insertion carrying larvae dies during the pupal stage, they grow properly without it [82].

Release of Insects Carrying a Dominant Lethal Gene (RIDL) is a concept that was introduced by the British biotech company Oxitec. Tetracycline, an antidote, can suppress the deadly gene so that mosquitoes can be raised to adulthood in breeding locations before being released into the wild as males, where they mate with wild females and produce offspring that die at the larval stage [83].

The first successful gene knockout and transgenesis experiment was performed on mosquitoes. Instinct biological control methods based on population modification of vector mosquitoes have shown incredible potential in the struggle against mosquito-borne illnesses. These strategies include the genetics based sterile insect technique (SIT), CRISPR/Cas9 mediated gene drive, and population-replacement methodologies [84].

The development of innovative nanotechnology-based formulations for natural and synthetic repellents is a necessary aspect towards more efficient techniques with fewer adverse side effects.

5. Conclusions

In recent years, a number of new diseases that are transmitted by mosquito vectors have evolved as a consequence of climate change, dramatic population increase, degradation and increased resistance to pesticides. However, the use of pesticides can have a negative impact on other forms of life, which can then lead to an imbalance in the ecosystem. As a result, it is of the utmost importance to focus on finding novel and effective strategies that are environmentally friendly, easy to handle, safe, and inexpensive with no negative impact on populations that are not being targeted. Control strategies based on chemical, biological, and physical elements have all been implemented to prevent the spread of mosquito-borne diseases as the conventional methods of regulating mosquito populations. Due to the innovation and latest research scientist are able to explore the non-conventional methods of mosquito management based on genetic modification, nanotechnology, etc. Harmonious utilization of various control methods is the best way to manage the mosquito population.

Acknowledgements

First of all, I am very thankful to the Almighty Allah who is the greatest creator of universe. He blessed and inspired me to complete my research work satisfactorily. I also pay my gratitude with heart and soul to the Holy Prophet (P.B.U.H) and his beloved and Holy Family. Throughout my research, I am deeply grateful to my most respected, gracious, knowledgeable, and reverend supervisor Dr. Muhammad Asrar Assistant Professor Department of Zoology for his consolidated and inspiring guidance.

Thanks

Thank you to my group members Zeeshan Yousaf, Usama Saleem, Muhammad Rashid, and Muhammad Faisal for their love, care, and friendship. I am very grateful my mother Balqees Anwar, aunt Khalida Akthar, and brother Usman Javed always raised their hands for my well-being.

May Allah bless all these bright minds with long, happy and peaceful lives including me (Ameen).

Author details

Zeeshan Javed[1], Saira Mansha[1], Usama Saleem[1], Asad Mangat[1], Bilal Rasool[1], Muhammad Imran[1], Amna Batool[1], Mashal Shahzadi[1], Tehreem Raza[1], Danish Riaz[2], and Muhammad Asrar[1*]

1 Department of Zoology, Government College University, Faisalabad, Pakistan

2 Department of Zoology, Division Science and Technology, University of Education, Township, Lahore, Pakistan

*Address all correspondence to: asrar_agri@yahoo.com

References

[1] Benelli G. Research in mosquito control: Current challenges for a brighter future. Parasitology Research. 2015;**114**(8):2801-2805

[2] Pimentel D, Zuniga R, Morrison D. Update on the environmental and economic costs associated with alien-invasive species in the United States. Ecological Economics. 2005;**52**(3):273-288

[3] Dahmana H, Mediannikov O. Mosquito-borne diseases emergence/resurgence and how to effectively control it biologically. Pathogens. 2020;**9**(4):310

[4] Mehlhorn H, Al-Rasheid KA, Al-Quraishy S, Abdel-Ghaffar F. Research and increase of expertise in arachno-entomology are urgently needed. Parasitology Research. 2012;**110**(1):259-265

[5] Benelli G. Research in mosquito control: Current challenges for a brighter future. Parasitology Research. 2015;**114**(8):2801-2805

[6] Kilpatrick AM. Globalization, land use, and the invasion of West Nile virus. Science. 2011;**334**(6054):323-327

[7] Hubálek Z, Halouzka J. West Nile fever—A reemerging mosquito-borne viral disease in Europe. Emerging Infectious Diseases. 1999;**5**(5):643

[8] Shukla D, Wijayapala S, Vankar PS. Effective mosquito repellent from plant-based formulation. International Journal of Mosquito Research. 2018;**5**(1):19-24

[9] Moyes CL, Vontas J, Martins AJ, Ng LC, Koou SY, Dusfour I, et al. Contemporary status of insecticide resistance in the major Aedes vectors of arboviruses infecting humans. PLoS Neglected Tropical Diseases. 2017;**11**(7):e0005625

[10] Wilke ABB, Marrelli MT. Genetic control of mosquitoes: Population suppression strategies. Revista do Instituto de Medicina Tropical de Sao Paulo. 2012;**54**:287-292

[11] Benelli G, Beier JC. Current vector control challenges in the fight against malaria. Acta Tropica. 2017;**174**:91-96

[12] Lok P, Dijk S. Malaria outbreak in Burundi reaches epidemic levels with 5.7 million infected this year. BMJ. 2019;**366**:15104

[13] Ramos-Castaneda J, Barreto dos Santos F, Martinez-Vega R, Galvão de Araujo JM, Joint G, Sarti E. Dengue in Latin America: Systematic review of molecular epidemiological trends. PLoS Neglected Tropical Diseases. 2017;**11**(1):e0005224

[14] Katzelnick LC, Coloma J, Harris E. Dengue: Knowledge gaps, unmet needs, and research priorities. The Lancet Infectious Diseases. 2017;**17**(3):e88-e100

[15] Villamil-Gómez WE, Rodríguez-Morales AJ, Uribe-García AM, González-Arismendy E, Castellanos JE, Calvo EP, et al. Zika, dengue, and chikungunya co-infection in a pregnant woman from Colombia. International Journal of Infectious Diseases. 2016;**51**:135-138

[16] Brady OJ, Hay SI. The first local cases of Zika virus in Europe. The Lancet. 2019;**394**(10213):1991-1992

[17] Diallo D, Sall AA, Diagne CT, Faye O, Faye O, Ba Y, et al. Zika virus emergence in mosquitoes in southeastern Senegal, 2011. PLoS One. 2014;**9**(10):e109442

[18] Musso D, Nilles EJ, Cao-Lormeau VM. Rapid spread of emerging Zika virus in the Pacific area.

Clinical Microbiology and Infection. 2014;**20**(10):O595-O596

[19] Edington F, Varjao D, Melo P. Incidence of articular pain and arthritis after chikungunya fever in the Americas: A systematic review of the literature and meta-analysis. Joint, Bone, Spine. 2018;**85**(6):669-678

[20] Silva NIO, Sacchetto L, de Rezende IM, Trindade GDS, LaBeaud AD, de Thoisy B, et al. Recent sylvatic yellow fever virus transmission in Brazil: The news from an old disease. Virology Journal. 2020;**17**(1):1-12

[21] Fernandes NCCDA, Cunha MS, Guerra JM, Réssio RA, Cirqueira CDS, D'Andretta Iglezias S, et al. Outbreak of yellow fever among nonhuman primates, Espirito Santo, Brazil, 2017. Emerging Infectious Diseases. 2017;**23**(12):2038-2041

[22] Auguste AJ, Lemey P, Bergren NA, Giambalvo D, Moncada M, Morón D, et al. Enzootic transmission of yellow fever virus, Venezuela. Emerging infectious diseases. 2015;**21**(1):99

[23] Selemane I. Epidemiological monitoring of the last outbreak of yellow fever in Brazil—An outlook from Portugal. Travel Medicine and Infectious Disease. 2019;**28**:46-51

[24] Fang Y, Zhang Y, Zhou ZB, Xia S, Shi WQ, Xue JB, et al. New strains of Japanese encephalitis virus circulating in Shanghai, China after a ten-year hiatus in local mosquito surveillance. Parasites & Vectors. 2019;**12**(1):1-14

[25] López-Ruiz N, del Carmen Montaño-Remacha M, Durán-Pla E, Pérez-Ruiz M, Navarro-Marí JM, Salamanca-Rivera C, et al. West Nile virus outbreak in humans and epidemiological surveillance, West Andalusia, Spain, 2016. Eurosurveillance. 2018;**23**(14):17-00261

[26] Diaz A, Coffey LL, Burkett-Cadena N, Day JF. Reemergence of St. Louis encephalitis virus in the Americas. Emerging Infectious Diseases. 2018;**24**(12):2150

[27] Guo WP, Tian JH, Lin XD, Ni XB, Chen XP, Liao Y, et al. Extensive genetic diversity of Rickettsiales bacteria in multiple mosquito species. Scientific Reports. 2016;**6**(1):1-11

[28] Krajacich BJ, Huestis DL, Dao A, Yaro AS, Diallo M, Krishna A, et al. Investigation of the seasonal microbiome of anopheles coluzzii mosquitoes in Mali. PLoS One. 2018;**13**(3):e0194899

[29] Dieme C, Bechah Y, Socolovschi C, Audoly G, Berenger JM, Faye O, et al. Transmission potential of rickettsia felis infection by Anopheles gambiae mosquitoes. Proceedings of the National Academy of Sciences. 2015;**112**(26):8088-8093

[30] Eliasson H, Broman T, Forsman M, Bäck E. Tularemia: Current epidemiology and disease management. Infectious Disease Clinics. 2006;**20**(2):289-311

[31] San Martín JL, Brathwaite O, Zambrano B, Solórzano JO, Bouckenooghe A, Dayan GH, et al. The epidemiology of dengue in the Americas over the last three decades: A worrisome reality. The American Journal of Tropical Medicine and Hygiene. 2010;**82**(1):128

[32] Girod R, Guidez A, Carinci R, Issaly J, Gaborit P, Ferrero E, et al. Detection of chikungunya virus circulation using sugar-baited traps during a major outbreak in French Guiana. PLoS Neglected Tropical Diseases. 2016;**10**(9):e0004876

[33] Dale P, Knight J. Mosquito Control: Perspectives on Current Issues and Challenges. Annals of Community Medicine and Practice; 2017;**3**(2):1023

[34] Bowman LR, Donegan S, McCall PJ. Is dengue vector control deficient in effectiveness or evidence? Systematic review and meta-analysis. PLoS Neglected Tropical Diseases. 2016;**10**(3):e0004551

[35] Bai L, Morton LC, Liu Q. Climate change and mosquito-borne diseases in China: A review. Globalization and Health. 2013;**9**(1):1-22

[36] Ferguson NM. Challenges and opportunities in controlling mosquito-borne infections. Nature. 2018;**559**(7715):490-497

[37] World Health Organization & UNICEF. 2017. Global vector control response 2017-2030

[38] World Health Organization. (2013). Larval Source Management: A Supplementary Malaria Vector Control Measure: An Operational Manual

[39] Allgeier S, Friedrich A, Brühl CA. Mosquito control based on bacillus thuringiensis israelensis (Bti) interrupts artificial wetland food chains. Science of the Total Environment. 2019;**686**:1173-1184

[40] Bruehl CA, Despres L, Frör O, Patil CD, Poulin B, Tetreau G, et al. Environmental and socioeconomic effects of mosquito control in Europe using the biocide bacillus thuringiensis subsp. israelensis (Bti). Science of the Total Environment. 2020;**724**:137800

[41] Wang GH, Gamez S, Raban RR, Marshall JM, Alphey L, Li M, et al. Combating mosquito-borne diseases using genetic control technologies. Nature Communications. 2021;**12**(1):1-12

[42] Jamrozik E, de la Fuente-Núñez V, Reis A, Ringwald P, Selgelid MJ. Ethical aspects of malaria control and research. Malaria Journal. 2015;**14**(1):1-7

[43] Kaura T, Walter NS, Kaur U, Sehgal R. Different Strategies for Mosquito Control. Challenges and Alternatives; Intechopen. 2022:104594

[44] Murugan K, Benelli G, Ayyappan S, Dinesh D, Panneerselvam C, Nicoletti M, et al. Toxicity of seaweed-synthesized silver nanoparticles against the filariasis vector Culex quinquefasciatus and its impact on predation efficiency of the cyclopoid crustacean Mesocyclops longisetus. Parasitology Research. 2015;**114**(6):2243-2253

[45] Acevedo MA, Prosper O, Lopiano K, Ruktanonchai N, Caughlin TT, Martcheva M, et al. Spatial heterogeneity, host movement and mosquito-borne disease transmission. PLoS One. 2015;**10**(6):e0127552

[46] Achee NL, Bangs MJ, Farlow R, Killeen GF, Lindsay S, Logan JG, et al. Spatial repellents: From discovery and development to evidence-based validation. Malaria Journal. 2012;**11**(1):1-9

[47] Syafruddin D, Asih PB, Rozi IE, Permana DH, Hidayati APN, Syahrani L, et al. Efficacy of a spatial repellent for control of malaria in Indonesia: A cluster-randomized controlled trial. The American Journal of Tropical Medicine and Hygiene. 2020;**103**(1):344

[48] Xue RD, Qualls WA, Smith ML, Gaines MK, Weaver JH, Debboun M. Field evaluation of the off! Clip-on mosquito repellent (metofluthrin) against Aedes albopictus and Aedes taeniorhynchus (Diptera: Culicidae) in northeastern Florida. Journal of Medical Entomology. 2012;**49**(3):652-655

[49] Shah G, Shri R, Panchal V, Sharma N, Singh B, Mann AS. Scientific basis for the therapeutic use of Cymbopogon citratus, stapf (lemon grass). Journal of Advanced Pharmaceutical Technology & Research. 2011;**2**(1):3

[50] Pichersky E, Gershenzon J. The formation and function of plant volatiles: Perfumes for pollinator attraction and defense. Current Opinion in Plant Biology. 2002;**5**(3):237-243

[51] Sugumar S, Mukherjee A, Chandrasekaran N. Eucalyptus oil nanoemulsion-impregnated chitosan film: Antibacterial effects against a clinical pathogen, Staphylococcus aureus, in vitro. International Journal of Nanomedicine. 2015;**10**(Suppl 1):67

[52] Tyagi BK. Advances in vector mosquito control technologies, with particular reference to herbal products. In: Herbal Insecticides, Repellents and Biomedicines: Effectiveness and Commercialization. New Delhi: Springer; 2016. pp. 1-9

[53] Chen YC, Wang CY, Teng HJ, Chen CF, Chang MC, Lu LC, et al. Comparison of the efficacy of CO2-baited and unbaited light traps, gravid traps, backpack aspirators, and sweep net collections for sampling mosquitoes infected with Japanese encephalitis virus. Journal of Vector Ecology. 2011;**36**(1):68-74

[54] Bellini R, Medici A, Puggioli A, Balestrino F, Carrieri M. Pilot field trials with Aedes albopictus irradiated sterile males in Italian urban areas. Journal of Medical Entomology. 2013;**50**(2):317-325

[55] Gupta VK, Jindal V. Biotechnological approaches for insect pest management. In: Integrated Pest Management. Academic Press; 2014:311-335

[56] Quesada-Moraga E, Carrasco-Díaz JA, Santiago-Álvarez C. Insecticidal and antifeedant activities of proteins secreted by entomopathogenic fungi against Spodoptera littoralis (Lep., Noctuidae). Journal of Applied Entomology. 2006;**130**(8):442-452

[57] Raveen R, Kamakshi KT, Deepa M, Arivoli S, Tennyson S. Larvicidal activity of Nerium oleander L.(Apocynaceae) flower extracts against Culex quinquefasciatus say (Diptera: Culicidae). International Journal of Mosquito Research. 2014;**1**(1):38-42

[58] Vais H, Williamson MS, Devonshire AL, Usherwood PNR. The molecular interactions of pyrethroid insecticides with insect and mammalian sodium channels. Pest Management Science. 2001;**57**(10):877-888

[59] Amelia-Yap ZH, Chen CD, Sofian-Azirun M, Low VL. Pyrethroid resistance in the dengue vector Aedes aegypti in Southeast Asia: Present situation and prospects for management. Parasites & Vectors. 2018;**11**(1):1-17

[60] Madarieta SK, Salarda A, Benabaye MRS, Tangle JR. Use of Permethrin-Treated Curtains for Control of Aedes aegypti, in the Philippines. 1999

[61] Suman DS, Wang Y, Bilgrami AL, Gaugler R. Ovicidal activity of three insect growth regulators against Aedes and Culex mosquitoes. Acta Tropica. 2013;**128**(1):103-109

[62] Vail PV, Coulson JR, Kauffman WC, Dix ME. History of biological control programs in the United States Department of Agriculture. American Entomologist. 2001;**47**(1):24-49

[63] El-Tarabily KA, El-Saadony MT, Alagawany M, Arif M, Batiha GE, Khafaga AF, et al. Using essential oils to overcome bacterial biofilm formation and their antimicrobial resistance. Saudi Journal of Biological Sciences. 2021;**28**(9):5145-5156

[64] El-Ashry RM, El-Saadony MT, El-Sobki AE, El-Tahan AM, Al-Otaibi S, El-Shehawi AM, et al. Biological silicon

nanoparticles maximize the efficiency of nematicides against biotic stress induced by Meloidogyne incognita in eggplant. Saudi Journal of Biological Sciences. 2022;**29**(2):920-932

[65] Das PK, Amalraj DD. Biological control of malaria vectors. The Indian Journal of Medical Research. 1997;**106**:174-197

[66] Coelho WMD, Coêlho JDCA, Bresciani KDS, Buzetti WAS. Biological control of Anopheles darlingi, Aedes aegypti and Culex quinquefasciatus larvae using shrimps. Parasite Epidemiology and Control. 2017;**2**(3):91-96

[67] Jacups SP, Rapley LP, Johnson PH, Benjamin S, Ritchie SA. Bacillus thuringiensis var. israelensis misting for control of Aedes in cryptic ground containers in North Queensland, Australia. The American Journal of Tropical Medicine and Hygiene. 2013;**88**(3):490

[68] Evans HC, Elliot SL, Barreto RW. Entomopathogenic fungi and their potential for the management of Aedes aegypti (Diptera: Culicidae) in the Americas. Memórias do Instituto Oswaldo Cruz. 2018;**113**:206-214

[69] Das MK, Rao MRK, Kulsreshtha AK. Native larvivorous fish diversity as a biological control agent against mosquito larvae in an endemic malarious region of Ranchi district in Jharkhand, India. Journal of Vector Borne Diseases. 2018;**55**(1):34

[70] Sarwar M. Reducing dengue fever through biological control of disease carrier Aedes mosquitoes (Diptera: Culicidae). International Journal of Preventive Medicine Research. 2015;**1**(3):161-166

[71] Mullen GR, Durden LA, editors. Medical and Veterinary Entomology. Academic press; 2009

[72] Focks DA. Toxorhynchites as biocontrol agents. Journal of the American Mosquito Control Association. 2007;**23**(sp2):118-127

[73] Shaalan EAS, Canyon DV, Muller R, Younes MWF, Abdel-Wahab H, Mansour AH. A mosquito predator survey in Townsville, Australia, and an assessment of Diplonychus sp. and Anisops sp. predatorial capacity against Culex annulirostris mosquito immatures. Journal of Vector Ecology. 2007;**32**(1):16-21

[74] Petersen JJ. Nematodes as biological control agents: Part I. Mermithidae. Advances in Parasitology. 1985;**24**:307-344

[75] Thiyagarajan P, Kumar PM, Kovendan K, Murugan K. Effect of medicinal plant and microbial insecticides for the sustainable mosquito vector control. Acta Biologica Indica. 2014;**3**:527-535

[76] Benelli G, Caselli A, Canale A. Nanoparticles for mosquito control: Challenges and constraints. Journal of King Saud University-Science. 2017;**29**(4):424-435

[77] Bisset J, Rodríguez MM, Fernández D. Selection of insensitive acetylcholinesterase as a resistance mechanism in Aedes aegypti (Diptera: Culicidae) from Santiago de Cuba. Journal of Medical Entomology. 2006;**43**(6):1185-1189

[78] Li Y, Zhang J, Chen D, Yang P, Jiang F, Wang X, et al. CRISPR/Cas9 in locusts: Successful establishment of an olfactory deficiency line by targeting the mutagenesis of an odorant receptor co-receptor (Orco). Insect Biochemistry and Molecular Biology. 2016;**79**:27-35

[79] Thomas DD, Donnelly CA, Wood RJ, Alphey LS. Insect population

control using a dominant, repressible, lethal genetic system. Science. 2000;**287**(5462):2474-2476

[80] Wang Y, He X, Qiao L, Yu Z, Chen B, He Z. CRISPR/Cas9 mediates efficient site-specific mutagenesis of the odorant receptor co-receptor (Orco) in the malaria vector Anopheles sinensis. Pest Management Science. 2022;**78**(8):3294-3304

[81] Phuc HK, Andreasen MH, Burton RS, Vass C, Epton MJ, Pape G, et al. Late-acting dominant lethal genetic systems and mosquito control. BMC Biology. 2007;**5**(1):1-11

[82] Carvalho DO, McKemey AR, Garziera L, Lacroix R, Donnelly CA, Alphey L, et al. Suppression of a field population of Aedes aegypti in Brazil by sustained release of transgenic male mosquitoes. PLoS Neglected Tropical Diseases. 2015;**9**(7):e0003864

[83] Lacroix R, Delatte H, Hue T, Reiter P. Dispersal and survival of male and female Aedes albopictus (Diptera: Culicidae) on Reunion Island. Journal of Medical Entomology. 2009;**46**(5):1117-1124

[84] Nolan T. Control of malaria-transmitting mosquitoes using gene drives. Philosophical Transactions of the Royal Society B. 2021;**376**(1818):20190803

Chapter 3

Advances in Mosquito Control: A Comprehensive Review

Sarita Kumar and Arunima Sahgal

Abstract

Mosquitoes are the important global vectors transmitting diseases of human concern such as dengue, Chikungunya, Malaria, encephalitis and yellow fever, etc. Management of mosquito-borne diseases largely relies on the vector management because of the lack of effective medication and vaccination. Several strategies have been formulated and applied in the fields to control mosquitoes; yet there is a continued rise in mosquito-borne diseases leading to sufferings and morbidities. Presently, chemical interventions are the most preferred methods which has impacted human health and the environment negatively. These issues have created a demand to devise novel approaches which can be used safely and effectively for mosquito management. Thus, several innovative mosquito control interventions have been devised based on genetic, physical and behavioral modifications in mosquitoes. These strategies span from Sterile Insect Technique (SIT) Release of Insects Carrying a Dominant Lethal (RIDL), creating transgenics with abnormal and lethal genes, gene drive technology, reducing the vectorial capacity by *Wolbachia* infection and application of attractive toxic sugar baits (ATSB), or by lasers and light detectors to investigate their behavior, and enhance their trap and kill. This Chapter gives a comprehensive overview of the conventional, and novel and innovative techniques devised for the control of mosquito vectors.

Keywords: Mosquito control, SIT, *Wolbachia*, ATSB, Transgenics, Gene drive, RIDL, Laser, Conventional strategies

1. Introduction

Mosquitoes, *Aedes, Culex* and *Anopheles* sp., are the global vectors of public health importance. These are widely distributed throughout the world and are responsible for transmitting several diseases of human concern such as dengue, Chikungunya, Zika, Malaria, encephalitis and yellow fever, etc. The continuous rise in these diseases has created a worldwide concern. According to the World Health Organization [1], an estimated 241 million global cases of malaria were recorded in 2020 in comparison to the 227 million cases in 2019. Among these, India accounts for 3% of global malaria cases and 2% of malaria deaths across the globe [2]. Likewise, dengue cases have increased intensely leading to about 50% global population at risk of dengue transmission and approximately 100–400 million annual infections [3].

Aedes is a known vector of dengue, Chikungunya, Yellow fever and Zika. According to the reports, the most common urban species, *Ae. aegypti*, was originated as *Ae. aegypti formosus* in the wilds of Sub-Saharan Africa and gradually established globally, more specifically in the tropical and subtropical regions of the world [4–6]. Likewise, *Ae. albopictus*, the peri-urban and rural vector, a native zoophilic species of Southeast Asia, Western Pacific and islands of Indian Ocean, has expanded globally *via* human activities and active transportations [7–9]. It is believed that *Aedes* could spread and establish into new regions because of climatic changes, elevated carbon emissions and global warming leading to the global appearance and expansion of several *Aedes*-borne arboviral diseases [10–12]. *Aedes* is currently distributed throughout the tropics including Africa (from where it originates) and a number of subtropical regions such as South-Eastern United States, the Middle East, Southeast Asia, the Pacific and Indian Islands, and Northern Australia (**Figure 1**).

Among *Culex*, the *Cx. pipiens* is the most widely distributed species responsible for the transmission of encephalitis, West Nile Fever, St. Louis encephalitis, etc. The *Cx. pipiens pipiens*, an old word taxon, is prevalent in temperate regions dispersed from Northern Europe to the highlands of South Africa while *Cx. p. pallens* is distributed throughout temperate Asia and *Cx. p. fatigans* is prevalent in the tropical regions. Another common species, *Cx. quinquefasciatus* is present throughout the tropical and warm temperate regions (**Figure 2**). Presumed to be the native of lowlands of West Africa, it has spread to the New World *via* slave ships dispersing gradually to Asia and other tropical and temperate parts of the world [14].

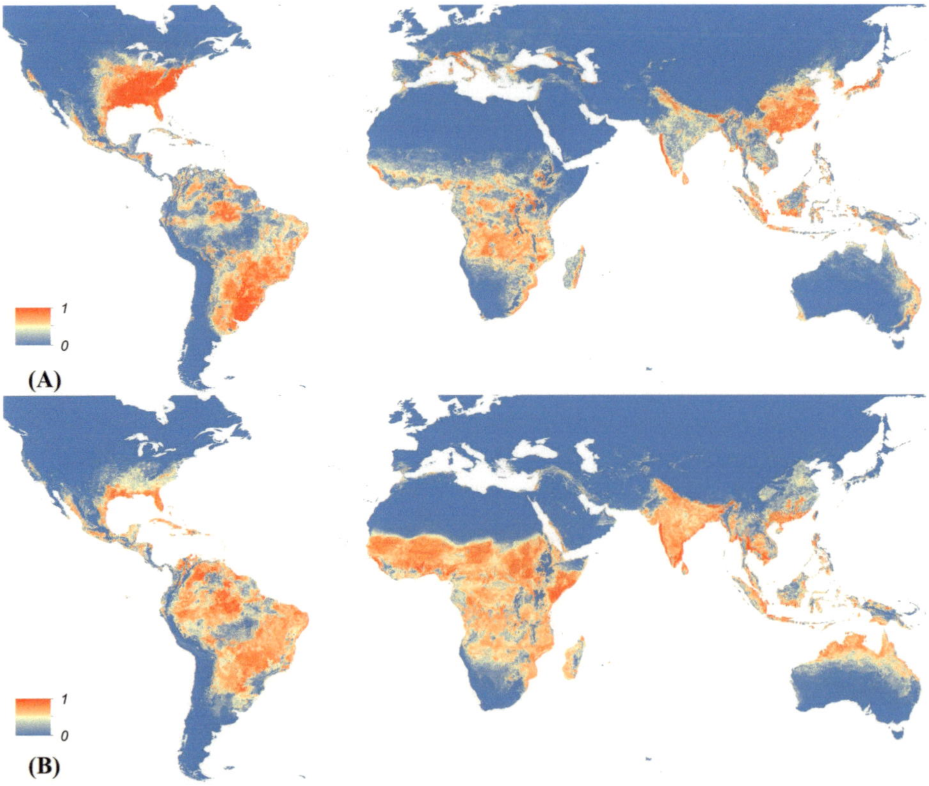

Figure 1.
Predicted distribution of Aedes aegypti (A) and Aedes albopictus (B) [6] (CC-BY-4.0).

DOI: http://dx.doi.org/10.5772/intechopen.106378

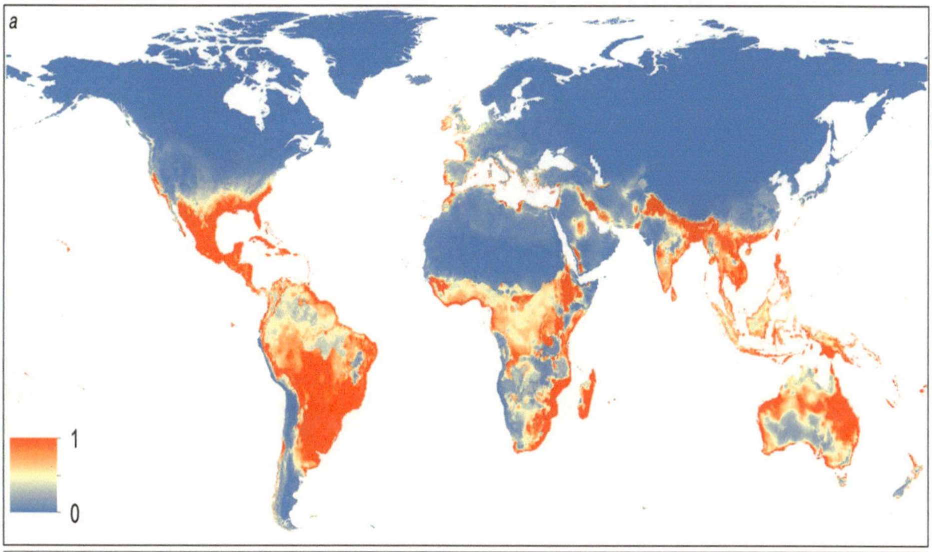

Figure 2.
Global distribution of Culex quinquefasciatus. Red areas represent the most suitable areas, yellow areas as potentially suitable while blue areas as unsuitable areas [13] (CC-BY-4.0).

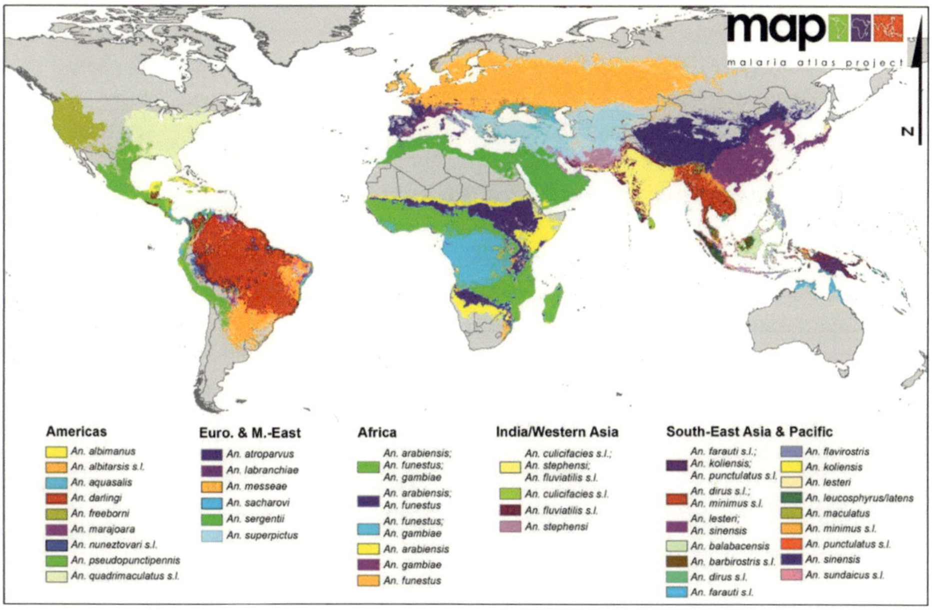

Figure 3.
Global distribution of Anopheles mosquito [15] (CC-BY-2.0).

Like other mosquito vectors, *Anopheles* species is also extensively distributed throughout the world, specifically in the tropical areas though with variability in the complexity (**Figure 3**). In Africa, the hardest hit, *An. gambiae, An. arabiensis* and *An. funestus* are the most common species, whereas in Indian region, *An. culicifacies, An. stephensi* and *An. fluviatilis* are the most prevalent. **Figure 4** depicts the status of

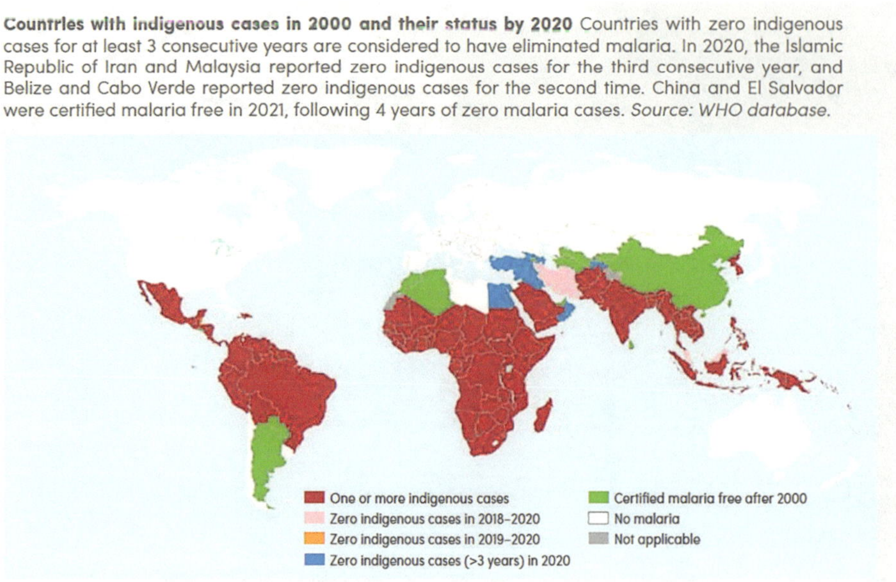

Figure 4.
Indigenous case status of malaria in 2020 [16].

malaria cases in these regions. China and El Salvador were declared malaria-free in 2021 by WHO as no malaria cases were reported in these countries for 4 consecutive years [16]. In addition, Islamic Republic of Iran and Malaysia have been reported to eliminate malaria with zero indigenous cases for 3 consecutive years (**Figure 4**).

2. Management of mosquitoes: Conventional strategies

Management of mosquito-borne diseases largely relies on the vector management because of the lack of effective medication and vaccination. Though, a few vaccines; such as dengvaxia is approved in some countries against dengue fever, 17D against yellow fever and RTS, S/AS01 against malaria; the use of these vaccines is associated with some constraints. For example, dengvaxia is considered effective only for 9–16 years old children and only when they have been previously infected with dengue and are living in areas where dengue is common [17]. Similarly, the WHO has endorsed RTS, S/AS01 vaccine against malaria but just for children and in regions with moderate to high *P. falciparum* malaria transmission [18].

Since olden times, several conventional strategies have been used to control mosquito vectors. Measures such as elimination of mosquito breeding sites and use of net screens on windows and doors to prevent entry of mosquitoes, etc. were commonly employed. During nineteenth century, the progress in science led to the formulation of DDT (an organochlorine) which was used as Indoor Residual Spray (IRS) during the Global Malaria Eradication Campaign (1955–1969) [19]. Gradually, other conventional synthetic insecticides – organophosphates, carbamates, pyrethroids; and Insect growth regulators, like JH analogues – Methoprene, Fenoxycarb, etc.; Chitin Synthesis Inhibitors – Dimiln, Penfluron, etc., were formulated which changed the direction of mosquito control [20]. These chemicals were used as active ingredients in the form of various formulations, sprays, dust, granules, and in mosquito repellents and bed nets, etc.

These interventions, devised for mosquito management, can be categorized into environmental interferences, chemical-based approaches, and biological control methods. Environmental management strategies include - sanitation, elimination of the mosquito breeding sites, avoid water stagnation, emptying water containers, covering all water-filled containers and waste management. Chemical interventions are based on the use of insecticides such as Temephos, Malathion and pyrethroids. Biological methods use agents such as copepods, larvivorous fish, *Bacillus sphaericus*, dragon fly naiads, may fly naiads, etc. A few measures to control *Aedes*-borne arboviral transmission have been depicted in **Figure 5**.

The use of insecticide-treated door curtains and bed nets (ITN), residual sprays in peri-domestic spaces, indoor residual spraying (IRS) and the control of larval breeding by Temephos and diflubenzuron, etc. are other commonly used approaches for mosquito management. In fact, use of IRS, ITNs (Insecticide-treated Nets) and LLINs (Long Lasting Insecticide-treated Nets) could reduce the malaria incidence in 21st century. Consequently, mass campaigns were held to distribute LLINs in countries with disease epidemics. However, the associated constrains; primarily development of insecticide resistance in mosquitoes and involved operational costs; limit the effectiveness of these approaches. Thus, efforts have been made to impregnate nets with synergized insecticides in order to reduce or reverse the resistance. Nevertheless. despite all efforts, the mosquito-borne diseases are continuing to rise in the world causing illness and morbidities at the global level.

Currently, majority of the mosquito control strategies are reliant on chemical-based interventions. However, use of these toxicants frequently and extensively has increased the problem of environmental pollution and led to widespread development of insecticide resistance in disease vectors. In addition, bioaccumulation of these chemicals in the environment has caused their biological magnification through the ecosystem. It is hypothesized that the vector control can be achieved fast with the implementation of new vector control interventions which can complement long-lasting insecticidal nets and indoor residual spraying. Thus, these issues have highlighted and necessitated the need to manage resistance, prevent resurgence of mosquito-borne diseases and maintain the drive towards disease elimination using biorational, effective and other novel approaches.

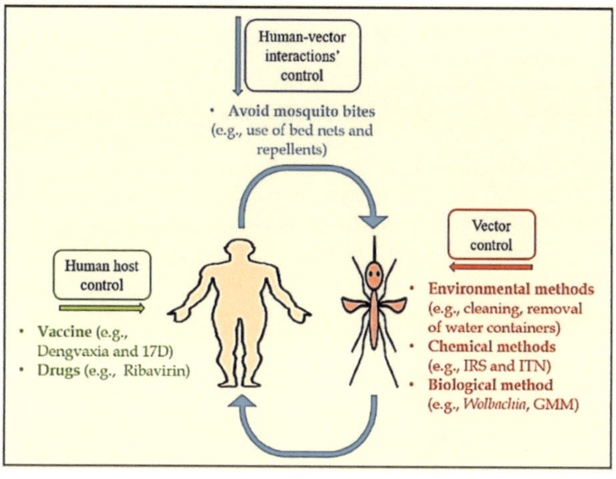

Figure 5.
Measures to control Aedes-borne arboviral transmission [21] (CC-BY-4.0).

3. Management of mosquitoes: Novel strategies

Successful control of mosquitoes requires a comprehensive approach. Since last few decades, several innovative mosquito control interventions and management approaches have been devised including genetic, physical and behavioral approaches. A brief and systematic review of these interventions has been discussed below.

3.1 Sterile insect technique

The use of sterile insects for insect pest management was first described in 1950's by E. F. Knipling. Though this innovative idea of paradigm shift in control interventions was initially used to eradicate screwworm flies, fruit flies etc., the use of SIT approach against mosquitoes has been conceived recently.

The SIT is a 4-step approach: mass production of mosquitoes, sorting males from females, irradiate male mosquitoes to make them sterile and mass release of sterile male mosquitoes into the target area (**Figure 6**). The objective behind the approach is that once released, sterile male mosquitoes compete with the wild males to mate with wild females. As SIT males are sterile, the mating does not produce any offspring and over time, the number of the targeted mosquito species in the area is reduced. The uniqueness of the SIT is that as the pest population reduces, the efficacy of the approach increases. Consequently, continued release of sterile mosquitoes reduces the vector population gradually over generations. Nonetheless, the approach can be successful in isolated population of mosquitoes and when the released male SIT mosquitoes are more numerous than the wild males.

The earlier attempts, however, had mixed success because it was highly problematic to rear and produce enough number of sterile males to suppress natural populations [23]. Further, sex separation in mosquitoes has been formerly based on the mechanical sorting according to the size dimorphism between male and female pupae

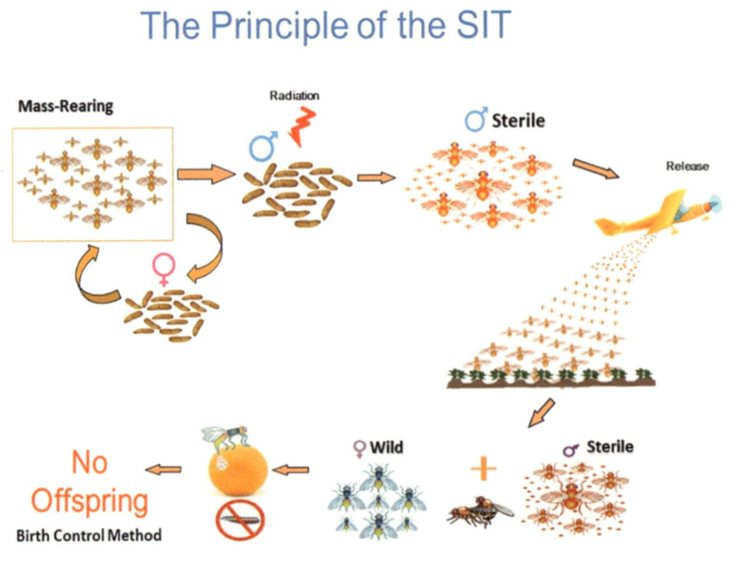

Figure 6.
Principle of sterile insect technique [22] (CC-BY-4.0).

as according to the reports, the pupal size-based sex separation could consistently give an essentially male-only population [24]. Nevertheless, irradiation at pupal stage used to often damage the insects in comparison to the irradiation at the adult stage which was, but, operationally much more difficult [25].

As a result, now, novel transgenic approaches have been created which help to develop and release required number of mosquitoes in the fields. Such approaches have been used for *An. gambiae* [26], *Ae. aegypti* [27], and *Ae. albopictus* [28]. Using these techniques, transgenic sterile *Ae. aegypti* have been released in the Grand Cayman and have demonstrated an effective reduction of these mosquitoes [29].

It is recommended that novel genetic methods using sex-linked markers may enhance the accuracy and efficiency of sex-sorting. For instance, the specificity of the homing endonuclease I-PpoI (Intron encoded endonuclease) of *An. gambiae* was exploited to distort sex ratio by producing only male offspring. The endonuclease selectively cleaved the ribosomal gene sequences located on the X chromosome (**Figure 7**). Slicing of the X chromosome prevents its transmission to the next generation and produced >95% male offspring [26].

The SIT technique is often used as a complement to other approaches as it is ecologically benign and insect-specific. The major limitation of SIT is that it is non-persistent in the environment because of inability of SIT mosquitoes to reproduce. Thus, once the release of these mosquitoes is stopped, the targeted mosquito species can return to normal. Hence, the technique demands regular release of irradiated males to be successful. Other constraints include – heavy expenditure to set up rearing and irradiation facility, complications in segregation of male mosquitoes, transportation issues, probable overdose of radiation which might affect vitals, and release of mosquitoes in isolated areas to avoid immigration of wild males and gravid females. The approach requires planning and commitment for long-term implementation due to its slow action as unlike fast-acting chemical interventions with immediate actions, it compromises the hereditary machinery of insect pest population by affecting next generation.

3.2 Release of insects carrying a dominant lethal (RIDL)

The limitations and issues associated with SIT could also be alleviated by the use of transgenic strains carrying specific novel traits, such as conditional genetic sterilization or lethality. The approach is based on the concept that transgenic strains of mosquitoes carrying a female-specific lethal gene could be used to remove females prior to release in the fields. It will remove the need for irradiation of males to manage vector population.

The RIDL approach acts late in the development and thus, it prevents mosquitoes from becoming adults. It was reported that introduction of LA 513 transposons into the mosquito's DNA produced offspring that die in the larval stage. The approach has been recommended in Malaysia as a control measure of *Ae. aegypti* [30].

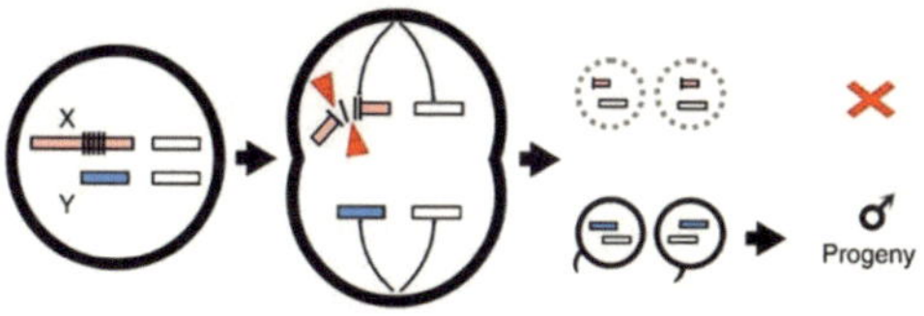

Figure 7.
Distortion of the sex ratio model in mosquitoes towards males based on meiotic X shredding [26] (CC-BY-NC-SA-3.0).

In another study, the "flightless female" *Ae. aegypti* has been developed by the genetic engineering of synthetic phenotypes [31]. The gene present in the strain encodes a toxin which destroys the wing muscles of females because of which the they are unable to fly, mate or search their food and oviposition sites. In *Ae. albopictus*, the gene *Actin-4* has been isolated to drive a dominant lethal gene in the indirect flight muscles leading to the development of a conditional female-specific flightless phenotype [28].

3.3 Male determining factor

The scientific advancement has discovered the male determining factor (M factor) in mosquitoes located within a Y chromosome–like highly repetitive M locus. These include *Nix* in *Ae. aegypti* [32], *gYG2/Yob* in *An. gambiae* [33, 34] and *Guy1* in *An. stephensi* [35]. A dominant male-determining locus (M-locus) establishes the male sex (M/m) in the yellow fever mosquito, *Ae. aegypti*.

It was discovered that knocking out the *Nix* gene results in feminized males while ectopic expression gave masculinized females with male genitalia [32]. Demonstration that M locus determines the male sex in *Ae. aegypti* and is thus inherited by only male mosquitoes has been carried out [36]. In case, the *Nix* gene inserted into a chromosomal region is inherited by the female *Ae. aegypti*, the mosquitoes can convert into non-biting males. They recommended that female-to-male sex conversion by *Nix* can complement SIT that requires only non-biting males and can help in reducing vector population to a great extent.

A *myo-sex* gene, need for flight, has also been discovered in the M-locus of male *Ae. aegypti* [36]. They demonstrated that the non-biting males converted from females lack this gene and thus were unable to fly for mating. They could not fold their wings completely but could walk and sometimes jump.

3.4 *Wolbachia*-based control

Wolbachia, an intracellular bacterium, is found in more than 50% of insect species and is transmitted vertically [37]. Mosquitoes do not possess *Wolbachia* but it can be introduced in the mosquito through trans-infections. The first *Wolbachia* strain (*wPip - Wolbachia pipientis*) was discovered in *Cx. pipiens* [38]. Later, other strains, *wAlbA* and *wAlbB* were found in *Ae. albopictus* [39].

The bacterium has been used to manage mosquito population, especially *Aedes*. It is reported that *Wolbachia*-based control techniques can not only disrupt replication and transmission of arbovirus; but the bacterium can also suppress the vector population [40]. *Wolbachia*-based mosquito control strategy involves two kinds of approaches – either replacement or suppression of the population [41]. In the population replacement strategy, the female mosquitoes infected with *Wolbachia* are released in the fields. The offspring of these mosquitoes are viable, whether they mate with *Wolbachia*-infected males or uninfected males. This allows the spread of *Wolbachia* in the field population. Consequently, though the total number of mosquitoes remains unchanged the individuals are less competent (**Figure 8**). On the other hand, in population suppression strategy, *Wolbachia*-infected males are released in the fields. These, when mate with the wild females, do not produce viable offspring. Thus, this strategy reduces the total number of mosquitoes instead of affecting their competency (**Figure 8**).

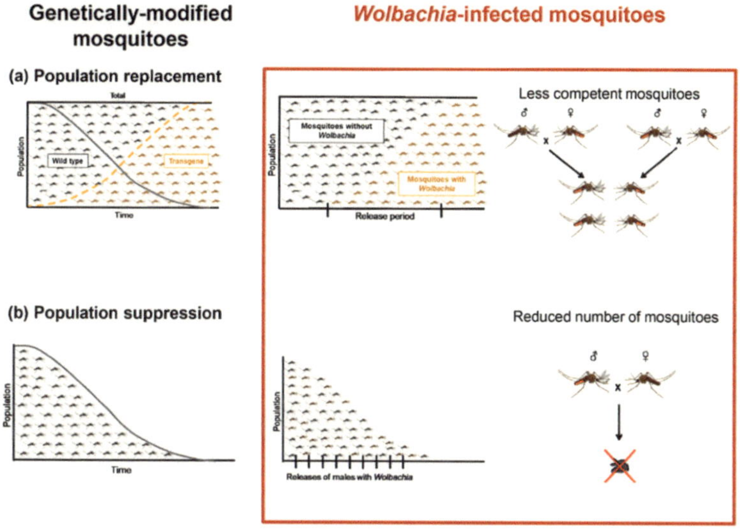

Figure 8.
Wolbachia-based mosquito control strategy [41] (CC-BY-4.0).

The studies have revealed that the *Wolbachia* trans-infection may decrease the fitness of *Aedes* leading to the population reduction and could halve its life-span [42, 43]. The study also showed that the mortality in *Wolbachia*-infected (*wMel, wAlbB, wMelPop*) mosquitoes was significantly higher in comparison to their wilds counterparts. In northern Australia, a *w*AlbB2-F4 strain has been generated which showed incompatibility with the wild strain as well as *w*Mel-*Wolbachia Ae. aegypti* [44]. The strain was mass reared and sexes were sorted in order to release only males in the field. They released 3 million males in 600 houses, approximately 50 males/house, 3 times a week for 20 weeks and recorded 80% decline in population in comparison to the control.

Research has shown that *Wolbachia*-infected strains have a competitive benefit over their wild counterparts. Mating between *Wolbachia*-infected male *Ae. aegypti* and wild female mosquitoes results in sterile eggs due to unidirectional cytoplasmic incompatibility [45]. In addition, the mating between male and female infected with different *Wolbachia* strains could also produce non-viable offspring (bidirectional incompatibility) [46]. Hence, in SIT, if *Wolbachia*-infected males are used and released regularly, the vector population can reduce drastically.

In addition, the *Wolbachia* has the capacity to block the transmission of dengue and Zika viruses by *Aedes* species. Studies have suggested that *Wolbachia*-infected *Ae. aegypti* might fix in the target population [47, 48]. In Kuala Lumpur, Malaysia, by 18 months of *Wolbachia*-host coevolution in the field, it was observed that blocking of dengue virus transmission and unidirectional cytoplasmic incompatibility were not compromised in a field-adapted *wAlbB*-carrying *Ae. aegypti* strain [49]. Thus, *Wolbachia*-based control can be three-pronged approach; release of *Wolbachia*-infected male mosquitoes along with uninfected female mosquitoes resulting in sterile offspring due to CI; introduction of *Wolbachia* strain causing fitness cost by reducing the life-span; and invasion of *Wolbachia* strain that inhibits virus transmission (**Figure 9**).

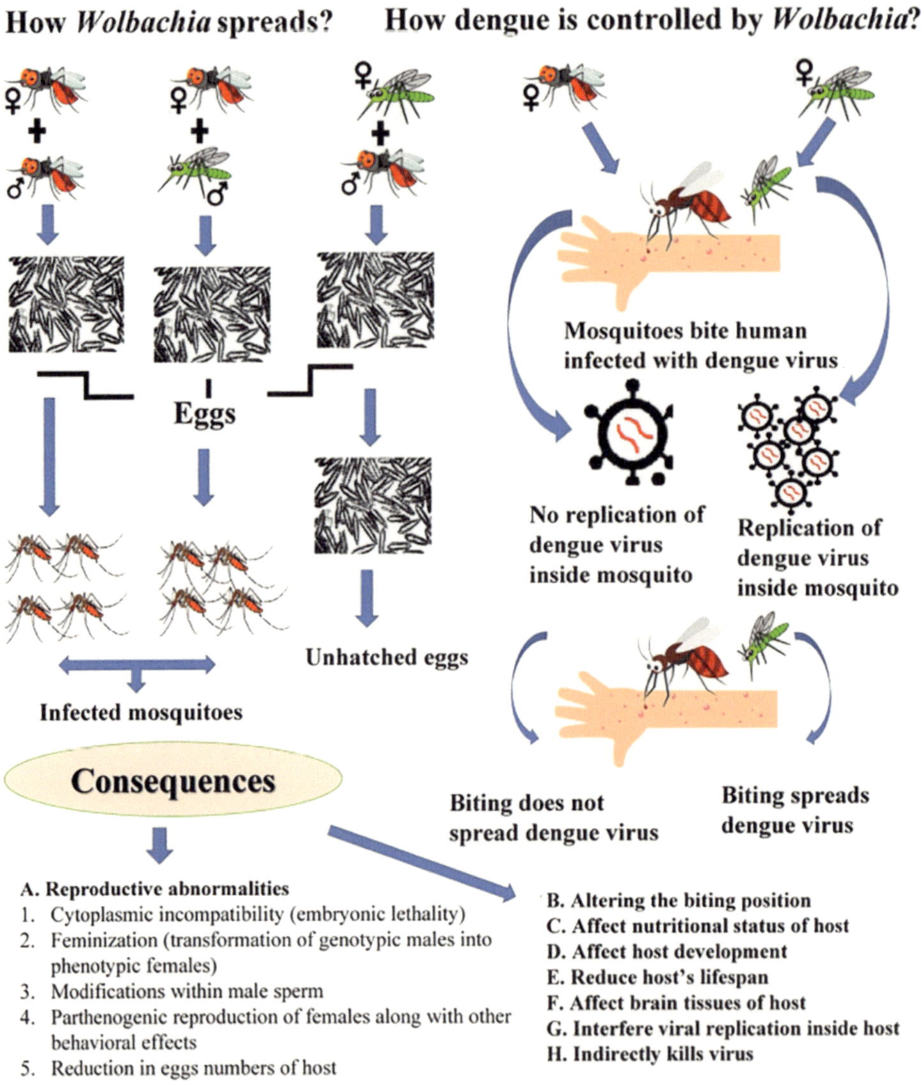

Figure 9.
Vertical transmission of Wolbachia and its role in reducing fitness of Aedes sp. and inhibition of dengue virus transmission. Red colored mosquitoes are infected with Wolbachia while green colored mosquitoes are uninfected [50] (CC-BY-4.0).

3.5 Gene drive

CRISPR-based gene drives are selfish genetic elements that can be used to modify entire populations of the mosquito for sustainable vector control [51]. Using gene drive technology, a genetic modification can spread through a population at higher inheritance rates than the normal. These technologies have been investigated in *An. gambiae, An. stephensi* and other mosquito species. The approach can either suppress a wild mosquito population or reduce its transmission competency by spreading genes that interfere with parasite development.

A highly effective and autonomous CRISPR-associated protein 9 (Cas9)-mediated gene-drive system has been developed in *An. stephensi* which resulted in progeny derived from transgenic males exhibiting a high frequency of germ-line gene conversion [52]. Earlier, it was used in *An. gambiae* for knocking out the genes responsible for female fertility exhibiting the potential to pass through consecutive generations [53].

3.6 Attractive toxic sugar bait (ATSB)

Attractive-toxic sugar baits are considered a new vector control paradigm based on "attract and kill" approach. The approach is based on the natural behavior of the mosquitoes to feed on plant sugars as an energy source immediately after emergence and intermittently during their life history. It is a known fact that the successful feeding by mosquito adults helps in high survival rate and reproductive fitness of mosquito. Though it is an old-age known fact, yet this behavior of mosquitoes was not tapped till recent times. Nowadays, the sugar feeding behavior of mosquito is being tapped to formulate ATSBs by combining a concentrated sugar-based food source, an olfaction stimulant to lure mosquitoes and a systemic insecticide to kill them. It is believed that development of ATSBs may contribute to their localized control.

ATSB approach has been favored as mosquito control strategy since use of a safe systemic toxin, such as boric acid, targets the sugar-seeking behavior of mosquitoes and can evade problems conventionally associated with the indiscriminate use of contact insecticides [54]. This method is suitable to be combined with any type of low-risk gut toxin, which makes it a potential and plausible tool to fight rising resistance against conventional contact pesticides [55].

Application of mosquito sugar-feeding behavior as a control strategy was first observed successfully in *Ae. aegypti* adults which when fed upon a paper incorporated with malathion-sugar solution exhibited 85.2% mortality [56]. Likewise, *Bacillus sphaericus* spores incorporated with the sucrose/dye solutions resulted in effective mortality in *Cx. pipiens* [57]. Now, researchers are exploring diverse toxic sugar baits containing fruit or flower juice as an attractant and different types and concentrations of toxic active ingredients against mosquito vectors.

The aim of using the attractant in the formulation is to manipulate mosquito behavior moving them away from their natural sugar sources, attracting towards bait and encourage feeding. The baits can be applied to the foliage, or kept in a simple trap, such as plastic bottles (bait stations) to lure and kill, capitalizing on resting and sugar seeking mosquitoes. The approach offers an exceptional tool of pesticide delivery. The concentrated sugar source prompts a natural feeding response in mosquitoes. The mosquitoes, then ingest the bait and receive the active toxic ingredients integrated with the bait [58]. The researchers are attempting to use contact insecticides too along with which can enter the mosquito body during their visit to the bait.

As mosquitoes spend most of their time in particular outdoor microhabitats, there is a high probability that mosquitoes ingest ATSB solution at their resting sites. Application of a dyed sucrose-spinosad solution on the tree flowers at desert areas had resulted in a substantial reduction in the feral mosquitoes as compared to that at the control locations [59]. The ATSB trials were held against *An. sergentii* and *An. gambiae* in an arid habitat of Israel with relatively little sugar source vegetation which proved to be highly successful [60]. The approach resulted in over 98% reduction in the sugar-poor sites. Interestingly, over 95% population reduced even in the sugar-rich sites. Likewise, a dramatic reduction of daily survival rates and malaria vectorial capacity was observed in *An. gambiae,* from 11.2 to 0.0 in sugar-poor sites and from 79.0 to 0.3 in sugar-rich sites. Similarly, about 90% of mosquitoes emerged from cisterns and wells were found to feed on ASB which suggested that applications of ATSB could be successful in storm drain systems [61].

A few successful ATSB field trials have also controlled *Cx. quinquefasciatus* from storm drains in Florida, USA and *An. gambiae* in Mali, West Africa reducing the population by 90% [62]. The ATSB treatments in Mali also reduced the longevity of older mosquitoes which had completed three or more gonotrophic cycles and decreased the mosquito prevalence from 37% pre-treatment to 6% post treatment in a month's time [63].

Application of a TSB containing 5% sucrose solution and 1% boric acid in the habitat of *Ae. taeniorhynchus* larvae could significantly reduce the landing rate counts of adults [64]. The application of 0.4% microencapsulated garlic-oil to local vegetation as the oral toxin component of ATSB for controlling *An. sergentii* populations inhabiting desert-surrounded wetlands in Israel reduced the biting-pressure by 97.5% [65]. Different juices and bait prototypes were investigated in laboratory and field-simulated conditions to control *An. arabiensis* [66]. The preference of mosquitoes was found for orange, watermelon and commercial guava juice over tomato, mango and banana. The results elucidated that the design of the sugar bait can influence feeding rates and, therefore, efficacy. Sugar baits that offered a resting surface were found more efficient and feeding on the sugar baits was maximized when these were placed close to peri-domestic vegetation.

In a laboratory study, 48 h exposure of *Ae. albopictus*, *Ae. taeniorhynchus* and *Cx. nigripalpus* to the non-flowering *Rhaphiolepis indica* (L.) plants sprayed with TSB containing 5% sucrose solution and 1% boric acid resulted in >96% mortality in all mosquito species [67]. Sugar baits formulated with boric acid, deltamethrin and dinotefuran tested against *Cx. quinquefasciatus* showed higher efficacy of boric acid and dinotefuran baits against resistant populations while that of deltamethrin bait against susceptible population [68]. The carbamate-resistant strain of *Cx. quinquefasciatus* has also been found significantly more affected by dinotefuran than the susceptible strain suggesting that toxicity of dinotefuran against mosquitoes is not strongly affected by the presence of common resistance mechanism, i.e., *kdr* mutation and insensitive acetylcholinesterase [69]. Indoor trials conducted with ATSB bait station, containing guava juice-bait mixed with chlorfenapyr (0.5%), boric acid (2%), oxazolamide (1%) in experimental huts (**Figure 10**) as well as window traps (**Figure 11**), could also

Figure 10.
ATSB station positioning in experimental huts [70] (CC-BY-4.0).

DOI: http://dx.doi.org/10.5772/intechopen.106378

Figure 11.
ATSB station positioning near windows [70] (CC-BY-4.0).

successfully attract and kill *Cx. quinquefasciatus* [70]. Use of aqueous sugar solutions in bait stations containing boric acid, fipronil, or spinosad resulted in significant reductions of local mosquito populations [59, 71].

Nevertheless, the selection of the toxicant to be included in an ATSB has remained a scientific concern as non-target species may be affected and also children may be attracted given the sweet nature of the substrate [72]. A comprehensive review of the advancements in the attractive toxic sugar baits for the mosquito control has highlighted their effects on the mosquito larvae and non-target insects, as well as has discussed future applications of ATSB methodologies [73]. The possible efficacy of TSBs and ATSBs in mosquito management recommends their integration into mosquito abatement programs.

3.7 Optical approaches

Various physical management strategies have been devised which use devices, such as microphones, sensors, lasers and light detectors to identify mosquitoes and detect, monitor, and investigate their behavior. The method aims to reduce energy consumption, and enhance trapping and killing of specific insects.

Lowell Wood, an astrophysicist invented a laser-based device to kill mosquitoes. The "Mosquito Laser" uses an imaging CCD Camera along with a LED and a retroflective material to detect mosquitoes. A high-power laser is then released on the detected target which kills it by physical disintegration. It is a very novel technique which is still in the prototype stage. However, this approach can be dangerous for humans as well as the environment.

The short (<25 ms) laser pulses have been used to kill/disable *An. stephensi* females (**Figure 12**). The researchers reported the higher efficacy of green and far-infrared

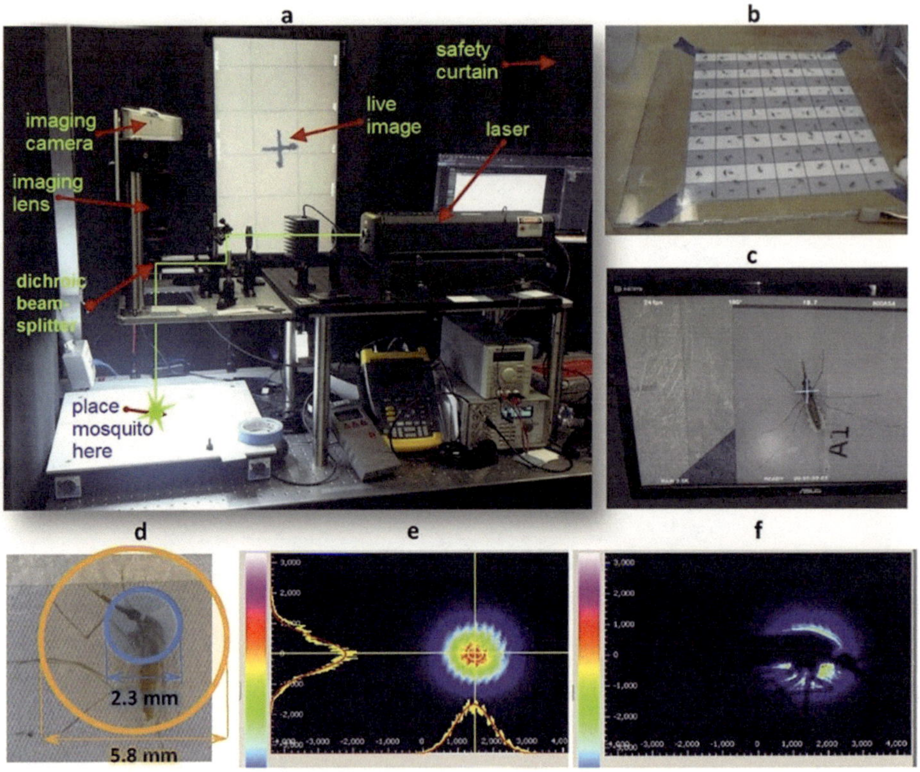

Figure 12.
Laser-induced killing of Anopheles stephensi (a) the current dosing laser co-aligned axially with an imaging camera; (b) anesthetized mosquitoes arranged in a 12 by 7 grid (c) view from the camera; (d) areas hit by the two typical beam diameters indicated by the circles; (e, f) Images from CCD beam (e) without and (f) with a subject [74] (CC-BY-4.0).

wavelengths in comparison to near- and mid-infrared wavelengths [74]. They recommended the use of cheap, robust lasers with sufficient beam quality so that they can be focused over a long range to kill mosquitoes. Earlier, efficacy of low power blue light to disable mosquitoes has been shown on exposure for several hours to days [75].

4. Conclusions

Prevention and control of mosquito-borne diseases, currently rely on the vector control due to the lack of effective medication and vaccines. The mosquito control has become a global challenge due to its widespread occurrence and transmission of diseases at a rapid rate. Despite the use of diverse conventional strategies; chemicals-based interventions, environmental management, human-vector interaction control, use of biological agents, etc.; the world is facing continual rise in these diseases. Thus, there is a need for the adequate implementation of these control strategies. Presently, mosquito control is based on the use of chemical insecticides. However, the recurrent and unsystematic use of these toxicants have caused harm to the human health, non-targets and the environment. Hence, researchers are incessantly exploring the alternate innovative strategies. The interventions which are covered in this Chapter can supplement the existing strategies and help to alleviate the mosquito population. Nevertheless, implementation of any new measure requires monitoring of its success which itself is a logistic challenge. It is recommended to carry out regular mosquito surveillance in the prevalent areas and implement the integrated mosquito control strategy.

Conflict of interest

The authors declare no conflict of interest.

Author details

Sarita Kumar[1]* and Arunima Sahgal[2]

1 Department of Zoology, Acharya Narendra Dev College (University of Delhi), Delhi, India

2 Department of Zoology, Ramjas College (University of Delhi), Delhi, India

*Address all correspondence to: saritakumar@andc.du.ac.in

References

[1] World Health Organization (WHO). Malaria. 2022a. Available from: https://www.who.int/news-room/fact-sheets/detail/malaria [Accessed: May 10, 2022]

[2] World Malaria Report. 2019. https://www.who.int/publications/i/item/9789241565721 [Accessed: May 2, 2022]

[3] World Health Organization (WHO). Dengue and Severe Dengue. 2022b. Available from: https://www.who.int/news-room/fact-sheets/detail/dengue-and-severe-dengue [Accessed: May 10, 2022]

[4] Brown JE, McBride CS, Johnson P, Ritchie S, Paupy C, Bossin H, et al. Worldwide patterns of genetic differentiation imply 'domestications' of *Aedes aegypti*, a major vector of human diseases. Proceedings of the Royal Society B: Biological Sciences. 2011;**278**(1717):2446-2454. DOI: 10.1098/rspb.2010.2469

[5] Brown JE, Evans BR, Zheng W, Obas V, Barrera-Martinez L, Egizi A, et al. Human impacts have shaped historical and recent evolution in *Aedes aegypti*, the dengue and yellow fever mosquito. Evolution. 2014;**68**(2):514-525. DOI: 10.1111/evo.12281

[6] Kraemer MU, Sinka ME, Duda KA, Mylne AQ, Shearer FM, Barker CM, et al. The global distribution of the arbovirus vectors *Aedes aegypti* and *Ae. albopictus*. eLife. 2015;**4**:e08347. DOI: 10.7554/eLife.08347

[7] Ponlawat A, Harrington LC. Blood feeding patterns of *Aedes aegypti* and *Aedes albopictus* in Thailand. Journal of Medical Entomology. 2005;**42**(5):844-849. DOI: 10.1093/jmedent/42.5.844

[8] Delatte H, Gimonneau G, Triboire A, Fontenille D. Influence of temperature on immature development, survival, longevity, fecundity, and gonotrophic cycles of *Aedes albopictus*, vector of chikungunya and dengue in the Indian Ocean. Journal of Medical Entomology. 2009;**46**(1):33-41. DOI: 10.1603/033.046.0105

[9] Izri A, Bitam I, Charrel RN. First entomological documentation of *Aedes* (*Stegomyia*) *albopictus* (Skuse, 1894) in Algeria. Clinical Microbiology and Infection. 2011;**17**(7):1116-1118. DOI: 10.1111/j.1469-0691.2010.03443.x

[10] Carrington LB, Simmons CP. Human to mosquito transmission of dengue viruses. Frontiers in Immunology. 2014;**5**:290. DOI: 10.3389/fimmu.2014.00290

[11] Semenza JC, Sudre B, Miniota J, Rossi M, Hu W, Kossowsky D, et al. International dispersal of dengue through air travel: Importation risk for Europe. PLoS Neglected Tropical Diseases. 2014;**8**(12):e3278. DOI: 10.1371/journal.pntd.0003278

[12] Messina JP, Brady OJ, Pigott DM, Golding N, Kraemer MU, Scott TW, et al. The many projected futures of dengue. Nature Reviews Microbiology. 2015;**13**(4):230-239. DOI: 10.1038/nrmicro3430

[13] Alaniz AJ, Carvajal MA, Bacigalupo A, Cattan PE. Global spatial assessment of *Aedes aegypti* and *Culex quinquefasciatus*: A scenario of Zika virus exposure. Epidemiology and Infection. 2019;**147**:E52. DOI: 10.1017/S0950268818003102

[14] CABI (Centre for Agriculture and Bioscience International). *Culex*

quinquefasciatus (Southern House mosquito). 2012. Available from: https://www.cabi.org/isc/datasheet/86848 [Accessed: May 17, 2022]

[15] Sinka ME, Bangs MJ, Manguin S, Rubio-Palis Y, Chareonviriyaphap T, Coetzee M, et al. A global map of dominant malaria vectors. Parasite & Vectors. 2012;**5**:69. DOI: 10.1186/1756-3305-5-69

[16] World Malaria Report 2021. Available from: https://www.who.int/publications/i/item/9789240040496 [Accessed: February 10, 2022]

[17] Centre for Disease Control and Prevention (CDC). Dengue vaccination: What everyone should Know. 2022. https://www.cdc.gov/vaccines/vpd/dengue/public/index.html [Accessed: May 11, 2022]

[18] World Health Organization (WHO). WHO recommends groundbreaking vaccine for children at risk. 2021. Available from: https://www.who.int/news/item/06-10-2021-who-recommends-groundbreaking-malaria-vaccine-for-children-at-risk [Accessed: May 11, 2022]

[19] Centre for Disease Control and Prevention (CDC). Indoor Residual Spraying. 2019. https://www.cdc.gov/malaria/malaria_worldwide/reduction/irs.html [Accessed: May 11, 2022]

[20] Mamatha DM, Kanji VK, Cohly HH, Rao MR. Juvenile hormone analogues, methoprene and fenoxycarb dose dependently enhance certain enzyme activities in the silkworm *Bombyx mori* (L). International Journal of Environmental Research and Public Health. 2008;**5**(2):120-124. DOI: 10.3390/ijerph5020120

[21] Ogunlade ST, Meehan MT, Adekunle AI, Rojas DP, Adegboye OA, McBryde ES. A review: *Aedes*-borne arboviral infections, controls and *Wolbachia*-based strategies. Vaccine. 2021;**9**(1):32. DOI: 10.3390/vaccines9010032

[22] Vreysen MJB, Abd-Alla AMM, Bourtzis K, Bouyer J, Caceres C, de Beer C, et al. The insect pest control laboratory of the joint FAO/IAEA programme: Ten years (2010-2020) of research and development, achievements and challenges in support of the sterile insect technique. Insects. 2021;**12**:346. DOI: 10.3390/insects12040346

[23] Benedict MQ, Robinson AS. The first releases of transgenic mosquitoes: An argument for the sterile insect technique. Trends in Parasitology. 2003;**19**(8):349-355. DOI: 10.1016/s1471-4922(03)00144-2

[24] Ansari MA, Singh KR, Brooks GD, Malhotra PR, Vaidyanathan V. The development of procedures and techniques for mass rearing of *Aedes aegypti*. Indian Journal of Medical Research. 1977;**65**(Suppl):91-99

[25] Andreasen MH, Curtis CF. Optimal life stage for radiation sterilization of *Anopheles* for sterile insect releases. Medical Veterinary Entomology. 2005;**19**:238-244. DOI: 10.1111/j.1365-2915.2005.00565.x

[26] Galizi R, Doyle L, Menichelli M, Bernardini F, Deredec A, Burt A, et al. A synthetic sex ratio distortion system for the control of the human malaria mosquito. Nature Communications. 2014;**5**:3977. DOI: 10.1038/ncomms4977

[27] Phuc HK, Andreasen MH, Burton RS, Vass C, Epton MJ, Pape G, et al. Late-acting dominant lethal genetic systems and mosquito control. BMC Biology. 2007;**5**:11. DOI: 10.1186/1741-7007-5-11

[28] Labbé GM, Scaife S, Morgan SA, Curtis ZH, Alphey L. Female-specific flightless (fsRIDL) phenotype for control of *Aedes albopictus*. PLoS Neglected Tropical Diseases. 2012;**6**:e1724. DOI: 10.1371/journal.pntd.0001724

[29] Harris AF, McKemey AR, Nimmo D, Curtis Z, Black I, Morgan SA, et al. Successful suppression of a field mosquito population by sustained release of engineered male mosquitoes. Nature Biotechnology. 2012;**30**:828-830

[30] Aldridge S. Genetically modified mosquitoes. Nature Biotechnology. 2008;**26**:828-830

[31] Macias VM, Ohm JR, Rasgon JL. Gene drive for mosquito control: Where did it come from and where are we headed? International Journal of Environmental Research and Public Health. 2017;**14**(9):1006. DOI: 10.3390/ijerph14091006

[32] Hall AB, Basu S, Jiang X, Qi Y, Timoshevskiy VA, Biedler JK, et al. A male-determining factor in the mosquito *Aedes aegypti*. Science. 2015;**348**:1268-1270

[33] Hall AB, Papathanos P-A, Sharma A, Cheng C, Akbari CC, Assour L, et al. Radical remodeling of the Y chromosome in a recent radiation of malaria mosquitoes. Proceedings of National Academy of Sciences USA. 2016;**113**:E2114-E2123. DOI: 10.1073/pnas.1525164113

[34] Krzywinska E, Dennison NJ, Lycett GJ, Krzywinski J. A maleness gene in the malaria mosquito *Anopheles gambiae*. Science. 2016;**353**:67-69. DOI: 10.1126/science.aaf5605

[35] Criscione F, Qi Y, Tu Z. GUY1 confers complete female lethality and is a strong candidate for a male-determining factor in *Anopheles stephensi*. eLife. 2016;**5**:e19281. DOI: 10.7554/eLife.19281

[36] Aryan A, Anderson MAE. Biedler Jk, Qi Y, Overcash JM, Naumenko AN, Sharakhova MV, Mao C, Adelman ZN, Tu Z. *nix* alone is sufficient to convert female *Aedes aegypti* into fertile males and *myo-sex* is needed for male flight. Proceedings of National Academy of Sciences USA. 2020;**117**(30):17702-17709. DOI: 10.1073/pnas.2001132117

[37] Hilgenboecker K, Hammerstein P, Schlattmann P, Telschow A, Werren JH. How many species are infected with *Wolbachia*? – A statistical analysis of current data. FEMS Microbiology Letters. 2008;**281**:215-220. DOI: 10.1111/j.1574-6968.2008.01110.x

[38] Hertig M, Wolbach S. Studies on rickettsia-like micro-organisms in insects. Journal of Medical Research. 1924;**44**:329-374

[39] Zhou W, Rousset F, O'Neill SL. Phylogeny and PCR-based classification of *Wolbachia* strains using *wsp* gene sequences. Proceedings of the Royal Society B: Biological Science. 1998;**265**:509-515. DOI: 10.1098/rspb.1998.0324

[40] Hoffmann AA, Ross PA, Rašić G. *Wolbachia* strains for disease control: Ecological and evolutionary considerations. Evolutionary Applications. 2015;**8**:751-768. DOI: 10.1111/eva.12286

[41] Yen P-S, Failloux A-B. A review: *Wolbachia*-based population replacement for mosquito control shares common points with genetically modified control approaches. Pathogens. 2020;**9**(5):404. DOI: 10.3390/pathogens9050404

[42] McMeniman CJ, Lane RV, Cass BN, Fong AWC, Sidhu M, Wang Y-F, et al. Stable introduction of a life-shortening *Wolbachia* infection into the mosquito *Aedes aegypti*. Science. 2009;**323**:141-144. DOI: 10.1126/science.1165326

[43] Axford JK, Callahan AG, Hoffmann AA, Yeap HL, Ross PA. Fitness of wAlbB *Wolbachia* infection in *Aedes aegypti*: Parameter estimates in an outcrossed background and potential for population invasion. The American Journal of Tropical Medicine and Hygiene. 2016;**94**:507-516. DOI: 10.4269/ajtmh.15-0608

[44] Beebe NW, Pagendam D, Trewin BJ, Bomer A, Bradford M, Ford A, et al. Releasing incompatible males drives strong suppression across populations of wild and *Wolbachia*-carrying *Aedes aegypti* in Australia. Proceedings of National Academy of Sciences USA. 2021;**118**(41):e2106828118. DOI: 10.1073/pnas.2106828118

[45] Flor M, Hammerstein P, Telschow A. *Wolbachia*-induced unidirectional cytoplasmic incompatibility and the stability of infection polymorphism in parapatric host populations. Journal of Evolutionary Biology. 2007;**20**:696-706. DOI: 10.1111/j.1420-9101.2006.01252.x

[46] Zhong Y, Li Z-X. Bidirectional cytoplasmic incompatibility induced by cross-order transfection of *Wolbachia*: Implications for control of the host population. Microbial Ecology. 2014;**68**:463-471. DOI: 10.1007/s00248-014-0425-2

[47] Nazni WA, Hoffmann AA, Noor Afizah A, Cheong YL, Mancini MV, Golding N, et al. Establishment of *Wolbachia* strain wAlbB in Malaysian populations of *Aedes aegypti* for dengue control. Current Biology. 2019;**29**:4241-4248.e5. DOI: 10.1016/j.cub.2019.11.007

[48] Ryan PA, Turley AP, Wilson G, Hurst TP, Retzki K, Brown-Kenyon J, et al. Establishment of wMel *Wolbachia* in *Aedes aegypti* mosquitoes and reduction of local dengue transmission in Cairns and surrounding locations in northern Queensland, Australia. Gates Open Research. 2020;**3**:1547. DOI: 10.12688/gatesopenres.13061.2

[49] Ahmad NA, Mancini M-V, Ant TH, Martinez J, Kamarul GMR, Nazni WA, et al. *Wolbachia* strain wAlbB maintains high density and dengue inhibition following introduction into a field population of *Aedes aegypti*. Philosophical Transactions of the Royal Society B. 2021;**376**:20190809. DOI: 10.1098/rstb.2019.0809

[50] Khadka S, Proshad R, Thapa A, Acharya KP, Kormoker T. *Wolbachia*: A possible weapon for controlling dengue in Nepal. Tropical Medicine and Health. 2020;**48**:50. DOI: 10.1186/s41182-020-00237-4

[51] Hammond A, Pollegioni P, Persampieri T, North A, Minuz R, Trusso A, et al. Gene-drive suppression of mosquito populations in large cages as a bridge between lab and field. Nature Communications. 2021;**12**:4589. DOI: 10.1038/s41467-021-24790-6

[52] Gantz VM, Jasinskiene N, Tatarenkova O, Fazekas A, Macias VM, Bier E, et al. Highly efficient Cas9-mediated gene drive for population modification of the malaria vector mosquito *Anopheles stephensi*. Proceedings of National Academy of Sciences USA. 2015;**112**(49):E6736-E6743. DOI: 10.1073/pnas.1521077112

[53] Hammond A, Galizi R, Kyrou K, Simoni A, Siniscalchi C, Katsanos D, et al. A CRISPR-Cas9 gene drive system targeting female reproduction in the malaria mosquito vector *Anopheles gambiae*. Nature Biotechnology. 2016;**34**:78-83. DOI: 10.1038/nbt.3439

[54] Enayati A, Hemingway J. Malaria management: past, present, and

future. Annual Review of Entomology. 2010;**55**:569-591. DOI: 10.1146/annurev-ento-112408-085423

[55] Allan SA. Susceptibility of adult mosquitoes to insecticides in aqueous sucrose baits. Journal of Vector Ecology. 2011;**36**(1):59-67. DOI: 10.1111/j.1948-7134.2011.00141.x

[56] Lea AO. Sugar-baited insecticide residues against mosquitoes. Mosquito News. 1965;**25**:65-66

[57] Schlein Y, Pener H. Bait-fed adult *Culex pipiens* carry the larvicide *Bacillus sphaericus* to the larval habitat. Medical Veterinary Entomology. 1990;**4**:283-288. DOI: 10.1111/j.1365-2915.1990.tb00441.x

[58] Stone CM, Jackson BT, Foster WA. Effects of bed net use, female size, and plant abundance on the first meal choice (blood vs sugar) of the malaria mosquito *Anopheles gambiae*. Malaria Journal. 2012;**11**:3-17. DOI: 10.1186/1475-2875-11-3

[59] Müller G, Schlein Y. Sugar questing mosquitoes in arid areas gather on scarce blossoms that can be used for control. International Journal for Parasitology. 2006;**36**:1077-1080. DOI: 10.1016/j.ijpara.2006.06.008

[60] Beier JC, Muller GC, Gu W, Arheart KL, Schlein Y. Attractive toxic sugar bait (ATSB) methods decimate populations of *Anopheles* malaria vectors in arid environments regardless of the local availability of favoured sugar-source blossoms. Malaria Journal. 2012;**11**:31-37. DOI: 10.1186/1475-2875-11-31

[61] Qualls WA, Xue RD, Revay EE, Allan SA, Muller GC. Implications for operational control of adult mosquito production in cisterns and wells in St. Augustine, FL using attractive sugar baits. Acta Tropica. 2012;**124**:158-161. DOI: 10.1016/j.actatropica.2012.07.004

[62] Müller GC, Junnila A, Qualls WA, Revay EE, Kline DL, Allan SA, et al. Control of *Culex quinquefasciatus* in a storm drain system in Florida using attractive toxic sugar bait. Medical and Veterinary Entomology. 2010a;**24**:346-351. DOI: 10.1111/j.1365-2915.2010.00876.x

[63] Müller GC, Beier JC, Traore SF, Toure MB, Traore MM, Bah S, et al. Successful field trial of attractive toxic sugar bait (ATSB) plant-spraying methods against malaria vectors in the *Anopheles gambiae* complex in Mali. West Africa. Malaria Journal. 2010b;**9**:210-216. DOI: 10.1186/1475-2875-9-210

[64] Hossain TT, Fulcher A, Davidson C, Beier JC, Xue RD. Evaluation of boric acid sprayed on plants against salt marsh mosquitoes, *Aedes taeniorhynchus* (Diptera: Culicidae). Florida Entomologist. 2014;**97**:1865-1868. DOI: 10.1653/024.097.0469

[65] Revay EE, Müller GC, Qualls WA, Kline DL, Naranjo DP, Arheat KL, et al. Control of *Aedes albopictus* with attractive toxic sugar baits (ATSB) and potential impact on non-target organisms in St. Augustine, Florida. Parasitology Research. 2014;**113**:73-79. DOI: 10.1007/s00436-013-3628-4

[66] Tenywal FC, Kambagha A, Saddler A, Maia MF. The development of an ivermectin-based attractive toxic sugar bait (ATSB) to target *Anopheles arabiensis*. Malaria Journal. 2017;**16**:338-347. DOI: 10.1186/s12936-017-1994-6

[67] Xue RD, Ali A, Kline DL, Barnard DR. Application of boric acid baits to plant foliage for adult mosquito control. Journal of the American Mosquito Control Association. 2006;**22**:497-500. DOI: 10.2987/8756-971X(2006)22[497,AOBABT]2.0.CO;2

[68] Gu ZY, He J, Teng XD, Lan CJ, Shen RX, Wang YT, et al. Efficacy of orally toxic sugar baits against contact-insecticide resistant *Culex quinquefasciatus*. Acta Tropica. 2020;**202**:105256. DOI: 10.1016/j.actatropica.2019.105256

[69] Corbel V, Duchon S, Zaim M, Hougard JM. Dinotefuran: a potential neonicotinoid insecticide against resistant mosquitoes. Journal of Medical Entomology. 2004;**41**(4):712-717. DOI: 10.1603/0022-2585-41.4.712

[70] Stewart ZP, Oxborough RM, Tungu PK, Kirby MJ, Rowland MW, Irish SR. Indoor application of attractive toxic sugar bait (ATSB) in combination with mosquito nets for control of pyrethroid-resistant mosquitoes. PLoS One. 2013;**8**:e84168. DOI: 10.1371/journal.pone.0084168

[71] Xue RD, Ali A, Kline DL, Barnard DR. Field evaluation of boric acid- and fipronil-based bait stations against adult mosquitoes. Journal of the American Mosquito Control Association. 2008;**24**:415-418. DOI: 10.2987/5683.1

[72] See AS, Salleh AB, Bakar FA, Yusof NA, Abdulamir AS, Lee YH. Risk and health effect of boric acid. American Journal of Applied Sciences. 2010;**7**:620-627

[73] Fiorenzano JM, Koehler PG, Xue R-D. Attractive toxic sugar bait (ATSB) for control of mosquitoes and its impact on non-target organisms: A review. International Journal of Environmental Research and Public Health. 2017;**14**:398-410. DOI: 10.3390/ijerph14040398

[74] Keller MD, Leahy DJ, Norton BJ, Johanson R, Mullen ER, Marvit M, et al. Laser induced mortality of *Anopheles stephensi* mosquitoes. Scientific Reports. 2016;**6**:20936. DOI: 10.1038/srep20936

[75] Hori M, Shibuya K, Sato M, Saito Y. Lethal effects of short-wavelength visible light on insects. Scientific Reports. 2014;**4**:07383. DOI: 10.1038/srep07383

Section 3

Biology and Management of Dipteran Pests

Chapter 4

Harmful Diptera Pests in Garlic and Onion and Their Management

Pervin Erdogan and Zemran Mustafa

Abstract

Garlic (*Allium sativum*) is a hardy perennial member of the onion family presumably native to Central Asia; however, it has long been naturalized in southern Europe and throughout the world. Onion, on the other hand, is used all over the world, and its consumption depends mostly on the income level of consumers. It is an indispensable vegetable in the kitchen of many homes. Onions take third place in vegetable production after potato and tomato in Turkey. Mites, nematodes, and insect species cause damage to these plants, reducing considerably their yield. Among these pests, the most destructive are *Delia platura* Meigen (Diptera: Anthomyiidae) *and Delia antiqua* Meigen (Diptera: Anthomyiidae). The crop losses can sometimes reach up to 100%, depending on the crop and density of the pest. There are different methods to control these pests that vary by the pest type and the crop being applied. *D. platura* eat the contents of newly planted seeds, leaving empty seed shells and preventing germination. Also, *D. antiqua*. Young onions are particularly vulnerable. When the hide and bulb become entangled in the damaged plant, development stops, the plant turns yellow, and it breaks. Both pests are controlled using biological and chemical methods.

Keywords: onion, garlic, pests, biology, management

1. Introduction

Garlic (*Allium sativum*) is a hardy perennial onion family member most likely native to Central Asia, but it has been adopted in southern Europe and many other cuisines. It is one of the oldest cultivated vegetables used for its edible parts and as traditional medicine. As it is with onions, garlic is not used alone in cooking. It is one of the indispensable vegetables of the kitchens due to its taste and flavoring properties. Garlic teeth and leaves have appetizing, diuretic, antibacterial, respiratory and digestive tract antiseptic, and antithyroid effects. In addition, garlic has the properties of purifying bile, lowering blood sugar and lipids, cough suppressant, blood pressure lowering, anti-infective, and curative [1].

Onion is used all over the world, and its consumption depends mostly on the income level of consumers. It is an indispensable vegetable in the kitchen of many homes giving flavor and taste to the dishes. It regulates metabolism and strengthens the immune system against microbial diseases. It is known as a beneficial food for breastfeeding mothers due to its stimulating milk-production properties. In addition,

IntechOpen

onion is recommended to be included in the nutrition programs for children growth as it is an important energy source and a medicinal plant whose use dates back to ancient years [2].

There are different methods to control these pests that vary by the pest type and the crop being applied. *Delia antiqua* Meigen (Diptera: Anthomyiidae), *Delia platura* Meigen (Diptera: Anthomyiidae), *Thrips tabaci* L. (Thysanoptera: Thripidae), *Frankliniella occidentalis* (Pergande) (Thysanoptera: Thripidae), *Rhyzoglyphus* spp., (Acari: Acaridae), *Tyrophagus* spp. (Acari: Acaridae), *Aceria tulipae* (Keifer) (Acari: Eriophyoidea) reduce the production of garlic. Previous studies carried out in Turkey reported garlic pests as *Bactericera tremblayi* Wagner (Hemiptera: Psylloidea), *T. tabaci, F. occidentalis* (Thysanoptera: Thripidae), *Acrolepiopsis assectella* (Zeller) (Lepidoptera: Acrolepiidae), *Agriotes* spp. (Coleoptera: Elateridae), *Liriomyza* spp. (Diptera: Agromyzidae), and *Ditylenchus dipsaci* (Kühn) (Tylenchida: Anguinidae) in garlic cultivation areas [3]. However, in recent years, garlic producers have complained about side effects of the pest on garlic production. The pests have been common and caused significant loss of garlic yield. Some producers have even maintained that they did not get any yields. The most important of these are *D. antiqua* Meigen (Diptera: Anthomyiidae and *D. platura* Meigen (Diptera: Anthomyiidae). Product loss as high as 35% on average is caused when no control is applied. The crop loss may sometimes reach up to 100%, depending on the crop and density of the pests. There are different pest control methods that vary by pest type. Besides chemical control for *D. platura* and *D. antiqua*, environmentally friendly practices such as yellow sticky traps are utilized for successful control of the Diptera pests [4].

2. Harmful Diptera species in garlic and onion

2.1 Seedcorn maggots (*D. platura*)

Seedcorn maggot or the bean seed fly, *D. platura* is a small dipterous insect. This fly is almost identical to the cabbage maggot (*Delia radicum*) and onion maggot (*D. antiqua*). In previous studies, this fly was named *Hylemia platura*, *Hylemia cana, Hylemia cilicrura, and Hylemia similis* [5]. Although *D. platura* was first reported in Germany, it is now spread all over the world [6].

The adult is about 6 mm long, gray in color, its body is covered with black hairs, and the wings are held crossed on the abdomen at rest. The egg is white in color, 0.6–0.7 mm long, and 0.2 mm wide. The larva is ivory-colored, thin on the front, wide on the back, and 5–6 mm in length. Pupa is barrel-shaped (**Figure 1a–d**).

There is no mandatory diapause. After the adult emerges from the pupa, it feeds on flowers and other sweet substances and lays its eggs for 10–18 days. Usually, seedcorn is found in newly plowed and irrigated fields. *D. platura* lays its eggs singly or in groups. It also lays eggs on degraded organic material in soil cracks and especially in manure piles. The hatched larvae immediately go under the soil and begin to feed on organic materials. Meanwhile, the larva searches for germinating seeds in seed beds and quarries. It feeds on the stems and cotyledons of young seedlings. Total larval development time is 8–11 days depending on temperature. *D. platura* has three larval stages. When the larva completes its development, it leaves the plant it feeds on and becomes a pupa at a depth of 5–6 cm in the soil. Pupation period is 10 days at 20°C. A female lays 40–50 eggs in her lifetime. Male flies live 20–25 days, while females

Figure 1.
Delia platura, (a) adult (wikipedia), (b) egg (Kilic, T), (c) pupae (Kilic, T.), (d) larva, (Erdogan, P), (e, f) damage (Erdogan, P).

live 30–35 days. The development period from egg to adult is 35 days at 10°C and 16–17 days at 25°C. *D. platura* produces 3–4 generations per year [7].

The seedcorn maggot is a polyphagous pest that attacks over 40 different host plants [8]. *Allium* species have been reported to host *D. platura* [9, 10]. The larvae feed by opening galleries in the stems and newly formed cotyledons of seeds germinating in seed beds or quarries such as squash, melon, cucumber, beans, garlic, and other vegetables [11]. When the seedling emerges from the soil, only the cotyledons attract attention because the shoot tip is eaten. During the seedling stage, plants turn yellow, and dry. *D. platura* is the most destructive in garlic. Seedcorn maggots, according to Bessin [12], damage newly planted seeds by feeding on seed contents, often leaving empty seed shells and preventing germination (**Figure 1e** and **f**). Seedlings that germinate despite the damage are spindly and have few leaves, and they die before maturing. Seedcorn maggots can sometimes be found tunneling within stems and germinating seeds [13]. Early planting dates, heavy cover crops, and cool-wet weather all contribute to seedcorn maggot damage [12]. *D. platura* also feeds in garlic. *D. platura* is the most harmful species in garlic fields, and it was spread throughout the entire area with the rate of harmful infestation increasing to 41% in some areas in Turkey [14]. In Ecuador, seedling losses of more than 60% have been reported 2 weeks after sowing. Plants that survive are weaker and less resistant to subsequent pest and pathogen attacks [15].

2.1.1 Management

2.1.1.1 Cultural control

- Planting in highly organic soils, wet soils, or during wet weather periods should be avoided as much as possible, as this frequently leads to increased risks.

- Manure should be allowed to age and be thoroughly incorporated into the soil before using.

- During seeding, dragging a chain behind the planter removes the seed row moisture gradient, which attracts oviposition by adult flies. This practice reduces seedcorn maggot damage significantly.

- Sanitation practices such as removing flowering weeds from outlying areas to eliminate nearby food sources for adult flies and removing sweet-smelling substances reduce an area's attractiveness to these flies.

- There have been isolated reports of spider and bird predation on adults, as well as fungus diseases on larvae. However, none of these predators are thought to be effective at controlling seedcorn maggot populations.

- Because they spend so much of their life underground, seedcorn maggots do not appear to have many natural enemies [16].

Kessing and Mau [11] suggested reusing manure and thoroughly incorporating it into the soil before planting. Planting when the soil and weather conditions favor rapid germination is a preventive measure against seed corn maggot infestation. According to Basin [12] planting, shallow planting, higher seeding rates, a well-prepared seedbed, and turning the cover early are all preventative measures for seed corn maggots (which renders the field less attractive to egg laying by flies). In another study, it was discovered that no-till fields are less likely to have seedcorn maggot problems because germinating seeds alone do not attract large populations of egg-laying females [17].

2.1.1.2 Biological control

Seedcorn maggots do not appear to have many natural enemies because they spend much of their life cycle underground. Isolated incidences of predation by spiders, ants, and birds upon adults and of fungal diseases infecting larvae have been reported, but none of these predators or pathogens is considered significant in controlling the population of seedcorn maggots [16]. There are some studies on the biological control of *D. platura*. Ellis and Scatcherd [4] revealed that two nematode parasitic species, *Steinernema feltiae* and *S. krausseri*, can be used to control seed fly and onion fly. The same study determined that *S. feltiae* reduced numbers of bean seed fly larvae by about 50%. Moreover, onion seedlings treated with the entomopathogenic fungus *Beauveria bassiana* (strain ANT-03) based insecticides were less attacked by the pest [18]. Recently, entomopathogenic nematodes (EPNs) have been proposed as a potential control method for *Delia* species [19, 20].

2.1.1.3 Chemical control

In problematic areas, seed spraying should be performed or seed beds should be sprayed before planting seeds. If seedlings are infected after planting, spraying should be done to prevent new contamination [11]. Before planting, the seeds should be sprayed with drugs containing active substances Imidacloprid, Thiamethoxam, and Acetamiprid. If there is a need to spray during the green period, a chemical pesticide containing cyromazine as active ingredient can be applied.

2.2 Onion fly (*Delia antiqua*)

Onion maggot, *D. antiqua*, is a major underground agricultural pest, found throughout Asia, Europe, and North America due to its high and low temperature tolerance [21, 22].

Adult of *D. antiqua* is similar to adult houseflies but slightly smaller. The adult is 6 mm long and gray in color. There are five dark bands on the thorax of the adults. Wings of adults are transparent and unblemished; legs and antennae are black. *D. antiqua* eggs are matte, white in color, 1.5 mm long, and banana-shaped. There are 12 longitudinal lines on the egg [23]. Larva is white and legless. The head of the body is narrow and shaped like a carrot that widens toward the end. Mature larva is 6–8 mm. *D. antiqua* pupae are reddish, 6 mm long, and barrel-shaped (**Figure 2a–d**).

Adults begin to appear in mid-March and early April, depending on climatic conditions. Adults emerge gradually throughout the following month. Temperature affects the length of the preoviposition period. Flies kept at 25°C laid eggs in 10–24 days, with a 15-day average. Adults lay their eggs in groups of 10–20 or singly, usually in the place where they meet with the onion, on leaves, tuber bark, and cracks in the soil, within 10–20 days following emergence. Flies in captivity lay up to 123 eggs at the rate of 1–45 per day. Eggs hatch in 3–6 days, depending on temperature and proportional humidity. The emerging larvae enter from the junction of the onion and the beet and move toward the inside of the onion. The larval period is 15–20 days. The mature larvae pupate in the soil near the onion plant [24]. The pupal period is 10–15 days. In addition to the damage it causes by feeding on the plant tissue, the larva causes decay with various bacteria it carries. The damage is especially greater in young onions. In the damaged plant, the development stops, the plant turns yellow, and it breaks when the hide and bulb are caught. Since the shallots that come out of the onion seeds planted for the purpose of growing shallots are frequent, the damage is more common (**Figure 2e** and **f**) [25]. The seed value of shallots is lost, and they rot. Onions, garlic, and bulbous flowers are their hosts [26, 27].

Figure 2.
Delia antiqua, (a) adult (Pest and diseases image library, (b) egg (Pest and diseases image library), (c) larva (Rasbak), (d) pupa (Rasbak), (e, f) damage (wikipedia.org).

2.2.1 Management

2.2.1.1 Cultural control

There are numerous non-chemical approaches to *D. antiqua* management. Crop rotation, crop sanitation, delayed planting, protective barrier installation, biological control, and sterile insect technique are examples of these.

- Crop rotation can considerably reduce damage by *D. antiqua.*

- Crop sanitation, including the removal of cull and volunteer onions, proper disposal of culls and volunteers, and avoiding damaging bulbs in the field is an important facet of *D. antiqua* management.

- Delayed planting can help as *D. antiqua* flies preferentially oviposit on larger onions.

- While onion plant varieties show little susceptibility to *D. antiqua*, garlic varieties have greater sensitivity.

- Since *D. antiqua* oviposits on or at the base of onion plants, the use of physical barriers to exclude flies has been considered; row covers effectively reduce infestations of both *D. platura* [28–31].

2.2.1.2 Biological control

Pests of *Delia* spp. are vulnerable to a variety of natural enemies and biocontrol agents. Many (60–100) species of staphylinid and carabid ground beetles, generalists that feed on eggs and early instars, prey on *Delia* spp. [32]. In addition to feeding on eggs, some staphylinid beetles, such as *Aleochara bilineata* Gyllenhal (**Figure 3a**) and *A. bipustulata* Linnaeus, parasitize *Delia* pupae [32]. *Aphaereta pallipes* Say (**Figure 3b**) is a braconid fly, which has a diverse host range and successfully parasitizes *D. antiqua.* Moreover, other biocontrol agents of *Delia* include predators and parasitoids as well as entomopathogenic fungi (EPFs) and nematodes (EPNs). *Entomophthora muscae* (**Figure 3c**), *Metarhizium anisopliae* (**Figure 3d**), and *B. bassiana* (**Figure 3e**) have all been found to be capable of infecting and killing *Delia* sp. pests in EPF screenings [33–35]. Similarly, EPNs, including commercially (Nemaplus, Biotem, Larvanem, Capsanem, etc.) available species such as *S. feltiae* (**Figure 3f**) and *Heterorhabditis bacteriophora* (**Figure 3g**) infect *D. antiqua* [36].

2.2.1.3 Sterile insect technique

Sterile insect technique (SIT) is a pest management technique that involves rearing and sterilizing large numbers of a pest before releasing them into the wild population. Sterile males compete with fertile males for the opportunity to mate with fertile females, reducing the number of viable offspring [37].

2.2.1.4 Chemical management

Growers rely heavily on chemical management tactics due to the practical and economic limitations of non-chemical management tools. Currently, *Delia antiqua* is managed principally with insecticides applied as seed treatments and in-furrow applications of chlorpyrifos at planting [38]. Because flies move and disperse during the day, sprays targeting flies are unlikely to contact the pest and are not recommended for fly management, although some growers continue to use them [29].

2.3 Leaf miners (*Liriomyza* spp.)

Adults are 1–2 mm long, gray-black in color. Larvae are up to 3 mm, white yellow in color, and clear (**Figure 4a–d**). Adults are on all leaves of the plant, and larvae are gallery life miner adults and live 25–30 days. A female lays about 400 eggs between two epidermises of the leaf. Larvae that emerge from egg feed by opening various galleries in the leaf epidermis. The mature larva, which has completed its development, leaves itself the soil to become pupae [39].

Figure 3.
(a) Aphaereta pallipes, (b) Aleochara bilineata, (c) Entomophthora muscae, (d) Metarhizium anispliae, (e) Beauveria bassiana, (f) Steinernema feltiae, (g) Heterorhabditis bacteriophora (wikipedia.org).

Figure 4.
Liriomyza spp. (a) Adult, (b) egg, (c) larva, (d) pupa (Payne J.A.), (e, f) damage (wikivand.com).

The females inflict small wounds on leaves, feed on the sap that comes out of it, and cause cell disruption. Feeding causes loss of healthy leaf tissue, so that plant cannot capture enough sunlight and often becomes infected with disease. Plants often fail to grow or produce crops. Then damaged areas turn yellow and dry and the leaves fall (**Figure 4e** and **f**). They delay development in young plants and seedlings. They cause quality and yield loss. They are harmful to tomatoes, eggplant, peppers, beans, peas, broad beans, lettuce, zucchini, cucumbers, spinach, onions, and leeks [39].

2.3.1 Management

2.3.1.1 Cultural control

- The soil is plowed at a depth of 10 cm, and the pupae in the soil are destroyed.
- Contaminated plant waste should be disposed.

2.3.1.2 Biotechnical control

It can be used as yellow sticky traps to control adults of *Liriomyza* spp.

3. Conclusions

Garlic and onion foods have been known for their health benefits since ancient times and are frequently used for both health promotion and flavor purposes. The positive effects of garlic and onion also called a natural antibiotic among people are innumerable, and new benefits continue to be discovered every day. Both play a huge role in the health of people, such as reducing the risk of various types of cancer, improving mood, and maintaining skin and hair health. There are many pests that limit the production of these plants. For example, *D. platura, Delia antiqua, Liriomyza* spp., and *Ditylenchus dipsaci*. Especially, *D. platura, D. antiqua* cause a high rate of product loss when these two pests are not controlled. In order to prevent product loss caused by *D. platura* and *D. antiqua*, correct control methods must be applied. Onion and garlic farmers should be informed about these pests and their control.

Author details

Pervin Erdogan[1*] and Zemran Mustafa[2]

1 Faculty of Agricultural Sciences and Technology, Plant Protection Department, Sivas University of Science and Technology, Sivas, Turkey

2 Faculty of Agricultural Sciences and Technology, Plant Production and Technologies Department, Sivas University of Science and Technology, Sivas, Turkey

*Address all correspondence to: pervinerdogan@sivas.edu.tr

References

[1] Ones MG, Hughes J. Biosynthesis of the flavor precursors of onion and garlic. Journal of Experimental Botany. 2004;**55**(404):1903 18. DOI: 10.1093/jxb/erh138

[2] Onion Health Research. Available from: https://www.onions-usa.org/all-about-onions/onion-health-research [Internet]. 2022

[3] Kececi M, Aydemir M, Yasarakinci N, Erdogan P, Kilic T, Emekci U. Plant protection tecnical instraction. Ministry of Agriculture and Forestry. 2008;**3**:306-308

[4] Ellis SA, Scatcherd JE. Bean seed fly (*Delia platura*, *Delia florilega*) and onion fly (*Delia antiqua*) incidence in England and an evaluation of chemical and biological control options. Annals of Applied Biology. 2007;**151**(2):259-267. DOI: 10.1111/j.1744-7348.2007.00170.x

[5] Higley LG, Pedigo LP. Seedcorn maggot (Diptera: Anthomyiidae) population biology and aestivation in Central Iowa. Environmental Entomology. 1984;**13**:1436-1442. DOI: 10.1093/ee/13.5.1436

[6] Gesell S, Calvin D. Seed corn maggot as a pest of field corn. In: Entomological Notes. University Park, PA: Department of Entomology Fact Sheet, Penn State University; 2000. p. 2

[7] Guerra P, Clifford P, Keil B, Stevenson C, Diego M, Samaniego Peralta E, et al. Larval performance and adult attraction of *Delia platura* (Diptera: Anthomyiidae) in a native and an introduced crop. Journal Economic Entomology. 2017;**110**(1):186-191. DOI: 10.1093/jee/tow237

[8] Ristich SS. Biology of Seed-Corn Maggot, *Hylemya Cilicrura* (Rondani), and Preliminary Studies on its Control [thesis]. Ithaca, New York: Cornell University; 1950

[9] Griffiths GCD. Relative abundance of the root maggots *Delia radicum* (L.) and *D. floralis* (fallen) (Diptera: Anthomyiidae) as pests of canola in Alberta. Quaestiones Entomologicae. 1986;**22**:253-260

[10] Howard RJ, Garland JA, Seaman WL. Disease and Pests of Vegetable Crops in Canada. An Illustrated Compendium. Ottawa, Canada: The Canadian Phytopathological Society and The Entomological Society of Canada; 1994

[11] Kessing JLM, Mau RFL. Seed Corn Maggot, *Delia platura* (Meigen). Crop Knowledge Master. Honolulu, Hawaii (unpublished): Department of Entomology; 1991

[12] Bessin R. Corn: Cool soils favor damage by some insect pests. Kentucky Pest News. Number 98. 2004

[13] Funderburk JE, Pedigo LP, Berry EC. Seedcorn maggot (Diptera: Anthomyidae) emergence in conventional and reduced-tillage soybean systems in Iowa. Journal of Economic Entomology. 1983;**76**:131-134. DOI: 10.1093/jee/76.1.131

[14] Erdogan P. A new pest species *Delia platura* Meigen, (Diptera: Anthomyiidae) on garlic (*Allium sativum* L., Alliaceae) area in Kastamonu Province of Turkey. Journal of Agricultural Sciences. 2020;**26**:263-270. DOI: 10.15832/ankutbd.511355

[15] Lomas L, Mazon N, Rivera ME, Peralta E. Cuantificación del daño y

alternativas para el control de la mosca de la semilla (*Delia platura* Meigen) en el cultivo de chocho (*Lupinus mutabilis* Sweet), en Ecuador. 2012. Available from: http://andescdp.org/sites/default/files/wikis/wiki_cdp8/INIAP_GA.pdf

[16] Reid WJ. Biology of the Seed-Corn Maggot in the Coastal Plain of the South Atlantic States. Technical Bulletin No. 723. Washington, DC: United States Department of Agriculture; 1940

[17] Pope R. Pests of germinating corn and soybean. Iowa State University, Extension. 1998;**IC-480**:65-66

[18] Fortier A M. Evaluation de l'efficacite de biopesticides et pesticides a risque reduit pour le controle de la mouche des semis, *Delia platura,* dans l'oignon jaune. 2016. pp. 1-17. Available from: https://www.mapaq.gouv.qc.ca/SiteCollectionDocuments/Agroenvironnement/

[19] Chen S, Moes M. Susceptibility of cabbage root maggot, *Delia radicum*, to entomopathogenic nematodes (Steinernematidae and Heterorhabditidae). Nematologia Mediterranea. 2003;**31**:157-162

[20] Nielsen O. Susceptibility of *Delia radicum* to steinernematid nematodes. BioControl. 2003;**48**:431-446. DOI: 10.1023/A:1024759229315

[21] Ishikawa Y, Tsukada S, Matsumoto Y. Effect of temperature and photoperiod on the larval development and diapause induction in the onion fly, *Hylemya antiqua* Meigen (Diptera: Anthmyiidae). Applied Entomology and Zoology. 1987;**22**:610-616. DOI: 10.1303/aez.22.610

[22] Chen B, Li W, Feng G, He Z. Laboratory rearing, biological characteristics and diapause induction of the onion maggot, *Delia antiqua*. Journal of Chongqing Normal University. 2010;**27**:9-13

[23] Havukkala I, Harris MO, Miller JR. Onion fly egg distribution in soil: A new sampling method and the effect of two granular insecticides. Entomologia Experimentails Applicata. 1992;**63**(3):283-289. DOI: 10.1111/J1570-7458-1992.tb01585x

[24] Tanaka K, Watari Y. Adult eclosion timing of the onion fly, *Delia antiqua*, in response to daily cycles of temperature at different soil depths. Naturwissenschaften. 2003;**90**:76-79. DOI: 10.1007/s00114-002-0390-3

[25] Judd GJR, Borden JH. Influence of different habitats and mating on olfactory behavior of onion flies seeking successfully parasitizes ional hosts. Journal of Chemical Ecology. 1992;**18**:605-620. DOI: 10.1007/BF00987823

[26] Miles M. Studies of British anthomyiid flies. VI. The annual cycle of generations in some anthomyiid root flies. Bulletin Entomology Research. 1955;**46**:11-19. DOI: 10.1017/S0007485300030728

[27] Hahn DA, Denlinger DL. Meeting the energetic demands of insect diapause: Nutrient storage and utilization. Journal Insect Physiology. 2007;**53**:760-773. DOI: 10.1016/j.jinsphys.2007.03.018

[28] Hough-Goldstein JA, Hess KA, Cates SM. Group effect on Seedcorn maggot (Diptera: Anthomyiidae) mating behavior. Annals of Entomological of Society, America. 1987;**80**:520-523. DOI: 10.1093/aesa/80.4.520

[29] Finch S, Eckenrode CJ, Cadoux ME. Behavior of onion maggot (Diptera: Anthomyiidae) in commercial onion fields treated regularly with parathion

sprays. Journal Economic Entomology. 1986;**79**:107-113. DOI: 10.1093/jee/79.1.107

[30] Mowry TM. Ovipositional patterns and larval movement of *Delia antiqua* (Diptera: Anthomyiidae) on sprouted bulb and seedling onions. Journal Economic Entomology. 1993;**86**:1440-1445. DOI: 10.1093/jee/86.5.1440

[31] Harris CR, Svec HJ, Begg JA. Mass rearing of root maggots under controlled environmental conditions: Seed-corn maggot, *Hylemya cilicrura,* bean seed fly, H. liturata, *Euxesta notata* and *Chaetopsis* sp. Journal of Economic Entomology. 1966;**59**:407-410. DOI: doi.org/10.1093/jee/59.2.407

[32] Whistlecraft JW, Harris CR, Tolman JH, Tomlin AD. Mass-rearing technique for *Aleochara bilineata* (Coleoptera: Staphylinidae). Journal Economic Entomology. 1985;**78**:995-997. DOI: 10.1093/jee/78.4.995

[33] Harbi A, Abbes K, Elimem M, Lazheri H, Meganck K, Chermiti B. The seedcorn maggot Delia platura (Diptera: Anthomyiidae): An emerging pest of garlic crops in Tunisia. EPPO Bulletin. 2022;**52**(1):149-153. DOI: 10.1111/epp.12841

[34] Davidson G, Chandler D. Laboratory evaluation of entomopathogenic fungi against larvae and adults of onion maggot, *Delia antiqua* (Diptera: Anthomyiidae). Journal of Economic Entomology. 2005;**98**:1848-1855. DOI: 10.1093/jee/98.6.1848

[35] Majchrowicz I, Tadeusz JP, Pierre HR, Nguya KM. Effects of entomopathogenic and opportunistic fungi on *Delia antigua* (Diptera: Anthomyiidae) at low relative humidity. Environmental Entomology. 1990;**19**(4):1163-1167. DOI: 10.1093/ee/19.4.1163

[36] Yildirim E, Hoy CW. Cyromazine seed treatments to control onion maggot, *Delia antiqua*, on green onions. Journal of Economic Entomology. 2003;**96**:6. DOI: 10.1093/jee/96.5.1494

[37] Klassen W. Sterile Insect Technique Principles and Practice in Area-Wide Integrated Pest Management. The Netherlands: Taylor & Francis, CRC Press; 2005. ISBN-13 978-1-4020-4051-1 (e-book)

[38] Reiners S, Curtis PD, Helms M, McGrath MT, Nault BA, Seaman A. Cornell integrated crop and pest management guidelines for commercial vegetable production. In: N.Y. State Integrated Pest Management Progress. Geneva, NY: Cornell University; 2019

[39] Tekeda M, Kawai A, Mitsunaga T, Tsukazaki H, Yamashita K, Wako T. A novel method for evaluating the egg killing defenses and varietal resistance of the bunching onion against *Liriomyza chienensis* (Diptera: Agromyzidae) via the artificial inoculation of eggs. Applied Entomology and Zoology. 2020;**55**:93-103. DOI: 10.1007/s13355-019-00657-7

Chapter 5

Walnut Husk Fly (*Rhagoletis completa* Cresson), the Main Burden in the Production of Common Walnut (*Juglans regia* L.)

Aljaz Medic, Metka Hudina, Robert Veberic and Anita Solar

Abstract

The walnut husk fly (*Rhagoletis completa* Cresson) is the most important pest of walnuts (*Juglans regia* L.). It causes economically significant crop losses (up to 80% yield loss) in many growing regions, including the United States and most European walnut-producing countries. This chapter describes the impact of pest infestation on yield quantity and quality along with the current geographic distribution of the pest. Its bionomy and infestation symptoms are described in detail. An overview of monitoring and control methods used is also provided, and new methods that may prove useful for walnut husk fly control are listed. Monitoring the occurrence of the pest is the most important part of controlling the walnut husk fly, as only with an effective monitoring system can insecticides be applied at the appropriate time. Emphasis is placed on biotic protection and the possible role of phenolic compounds in cultivar resistance to walnut husk fly. Other control methods (non-chemical, mechanical, and biological control) are also gaining importance in pest management as more and more active substances in pesticides are regulated or phased out each year. Mechanical control methods are more or less only suitable for walnuts grown in the protected areas.

Keywords: biotic protection, phenolic compounds, cultivar resistance, yield quality, yield quantity, bionomy, control methods

1. Introduction

The walnut husk fly (*Rhagoletis completa* Cresson) belongs to the group of fruit flies (Diptera: Tephritidae), which are the most important pests worldwide [1]. Fruit flies are considered one of the most important pests in the world due to the large economic impact and strict quarantine restrictions imposed by several countries. Larval feeding and oviposition by females render fruits or vegetables unusable and inedible [2], and losses can be as high as 80% of the yield [1]. Apart from the economic level, the fruit fly is also important at the ecological level, as it can displace native species or compete with them for resources [2].

Rhagoletis is a genus of Tephritidae fruit flies, which includes about 60 species. The walnut husk fly is the most destructive pest of the genus *Rhagoletis*, along with the European cherry fruit fly (*Rhagoletis cerasi* Loew) and the apple maggot (*Rhagoletis pomonella* Walsh) [1]. Fruit flies of the genus *Rhagoletis* have a specific combination of wing patterns consisting of pointed black or yellow postocular setae, four or five transverse bands, and posteroapical lobe along the anal vein. The male specimens have a long, saber-like surstyli genital, whereas the female specimens have a terminalia with a short oviscape that forms a soft, desclerotized T-shaped area at the apex [3]. Flies of the genus *Rhagoletis* are distributed throughout North and South America, Europe, and parts of Eurasia [4]. *Rhagoletis* species are usually found in temperate, mesic environments where rainfall is abundant. Most *Rhagoletis* species are univoltine, and only a few are multivoltine, producing a small second generation. Most species spend 8 to 10 months in the soil under host trees, from where they emerge from pupae between May and August (Northern Hemisphere). The female specimens emerge before the males. After emerging, they mate within 1 to 2 weeks on or near the fruit of their host plants. Both sexes are opportunistic feeders as their foods include fruit juice, exudates from extrafloral nectaries, plant leachates, bird excretions, homopteran honeydew, and possibly yeasts and bacteria. Male specimens are territorial and wait on host fruits, which they defend against other males, for the arrival of female specimens to mate with them. *Rhagoletis* species do not exhibit true courtship, as there are no predictable or elaborate behaviors that lead to mating and mounting. To mate, males simply attempt to hump the back of females. Males attempt to mate with females that are ovipositioning. The females then individually deposit the eggs directly under or on the skin of the fruit. In addition to depositing the eggs, females also mark the fruit with pheromones that prevent further oviposition and thus competition for resources within the species. After the larvae hatch from the eggs, they begin to feed on the pulp of the fruit, causing the fruit to rot and become unmarketable and inedible. There are three larval stages in *Rhagoletis* species. The final larval stage usually leaves the damaged fruit and burrows into the soil, where it pupates. It is believed that adults can survive up to 1 month in the wild, while the duration of the stage from egg to pupation depends on temperature and usually lasts between 4 and 10 weeks [5–9].

1.1 Hosts, origin, and distribution

Walnut husk fly belongs to the *suavis* species group. It has been classified there along with *Rhagoletis suavis* Loew, *Rhagoletis juglandis* Cresson, *Rhagoletis boycei* Cresson, *Rhagoletis ramosae* Hernández-Ortiz, and *Rhagoletis zoqui* Bush [9, 10]. Ten plants have been identified as hosts for the walnut husk fly. English or Persian walnut (*Juglans regia* L.) is the most commercially important, while black walnut (*Juglans nigra* L.) is the native host and is of lesser commercial importance. Other hosts include little or Texas walnut (*Juglans microcarpa* Berlandier), Arizona walnut (*Juglans major* (Torrey) Heller), Nuevo León walnut (*Juglans hirsuta* Manning), Hinds' black walnut (*Juglans hindii* Rehder), California black walnut (*Juglans californica* S. Watson) [6, 9], and Mexican walnut (*Juglans mollis* Engelm) [11]. Walnut husk fly has also been reported to infrequently attack peach (*Prunus persica* L. Batsch) [12], and Midland or English hawthorn (*Crataegus laevigata* (Poir.) DC.) [13].

The walnut husk fly is native to North America, to be more precise Midwestern USA and northeastern Mexico [9, 10]. It was first described and characterized by Cresson in the late 1920s [14]. From its native region in the Midwestern United States (Iowa, Kansas, Oklahoma, Minnesota, and Texas), it gradually spread to the other regions. In 1922, it was reported in California [12], from where it spread to the southern areas of California and northward to Washington State [15] and later to southern British Columbia (Canada) [5]. In Mexico, however, it is still restricted to the northeast, particularly to the states of Tamaulipas, Nuevo León, and Coahuila. In Nuevo León, walnut husk fly is restricted to *Janirella hirsuta* in higher-elevation areas within the canyons of the Sierra Madre Oriental [16], whereas in Tamaulipas and Coahuila it is infesting *J. mollis* [11].

It is believed that walnut husk fly spread from North America to Europe *via* global trade routes. In 1988, the walnut husk fly was documented for the first time in Europe, in Switzerland in the Ticino region [1, 17]. Later, the fly subsequently spread to neighboring countries. In 1991, it was reported near Venice and in the Friuli region [18], from where it rapidly spread throughout Italy. By 1992, it was confirmed in Milano, Novara, Varese, Pavia, and Sondrio [19]. From Italy, the walnut husk fly then spread across walnut-growing areas in Italy, and in 1997, the walnut husk fly was documented for the first time in Slovenia near Nova Gorica [20]. Documented observations of the walnut husk fly then increased each year: Croatia (2004), France (2007), Austria and Germany (2008), Albania (2010), Hungary (2011), Bosnia and Herzegovina, and Spain (2013) [1, 21–23]. Lacking natural predators, the walnut husk fly is expected to spread to all walnut-growing areas in Europe and Asia in the next few years, severely affecting walnut production [1].

1.2 Biology

To determine the species morphologically, a binocular microscope is recommended for diagnosis. Magnification for adult specimens is x10 and for larvae and aculeus x200. Only adult specimens can be reliably identified, while identification of eggs, larvae, and pupae is not reliable. Morphological identification of adult specimens can be done according to the EPPO (European and Mediterranean Plant Protection Organization) protocol [24], and that of larvae according to the protocol for the identification of larvae (third instar stage) [25]. The walnut husk fly life cycle, along with the damage caused can be seen in **Figure 1**.

Similar to other species of Tephritidae, the adult specimen of walnut husk fly is small, reaching a length of 4.0–6.5 mm. Both female and male specimens are of about the same size. The head of the walnut husk fly is completely yellow. It has a yellow dot on the posterior base of the thorax. Its wings are transparent and have three characteristic dark brown stripes. The last stripe is an elongated L that begins at the leading edge of the wing. Female specimens can be distinguished from males by a pointed abdomen with an ovipositor and by the color of the first leg segment. On males, the first leg segment is brown to black, whereas on females, it is pale yellow [9, 24]. A detailed morphology to distinguish *Rhagoletis* species is available at EPPO [24].

The walnut husk fly is univoltine (has only one generation per year). It usually overwinters in the soil under or near the host tree, as a pupa, however in years of high population density, individual specimens migrate elsewhere. Passive dispersal (wind, transport vehicles, etc.) is also likely. The life span of adult specimens is similar to other *Rhagoletis* species and is up to 40 days. Adult specimens emerge from the ground from mid-July to late September. Peak emergence is between late July and late August. In the host tree, they usually stay in the shaded part of the canopy where there are plenty of fruit. As previously seen with other *Rhagoletis* species, male specimens are territorial. They wait on the host fruit, for the arrival of the female specimens to mate with them. Mating occurs within 6 to 8 days after emergence from the soil. Females begin laying eggs 10 days after emerging from the soil.

Figure 1.
*Adult specimen of walnut husk fly (A), early signs of larvae damage (early: Bottom left, late: Bottom right, uninfested fruit: Above) (B), larvae (C), life cycle of walnut husk fly (D), and damage to shells and kernels (healthy left and damaged right) (E). *Photo of pupae provided by AGES/A. Egartner.*

A female can lay between 300 and 400 eggs in her lifetime. They usually lay eggs in groups of 15 to 20 per fruit. As with other *Rhagoletis* species, female specimens release a pheromone after laying eggs that prevents other females from laying eggs on the same fruit (walnut husk), which thus reduce larval competition for resources. One female can infest up to 20 fruits per season. After the females lay eggs on the fruit, the larvae hatch from the eggs in 3 to 10 days, depending on climatic conditions, especially temperature. The larvae are dirty white in color and have neither head nor legs. They reach a length of 8–10 mm. Once the larvae hatch, they bore into the walnut husk, where they make burrows and feed on the flashy pericarp (inner husk). Larval development takes between 30 and 40 days, during which they strip twice. Once they reach the last strip and are fully mature, they drop out of the husk onto the ground. They berry into the soil and pupate at a depth of a few centimeters. The pupation allows them to overwinter. The following year, more than 90% of the adults emerge from the soil and repeat this cycle again, while less than 10% of the pupae spend another season in diapause [1, 9, 26–29].

1.3 Economic damage

Economic damage is caused by the larvae of the walnut husk fly when they feed on the pericarp. This softens and blackens the husk, making it soft, sticky, and black. Once the damaged husk dries, it sticks to the endocarp (maturing nut) and is very difficult to remove or wash off. In severe infestations, the larvae may completely destroy the pericarp, leaving only the withered black exocarp around the shell [1]. Infestation causes the release of tannins from the damaged husks, resulting in black spots on the shells and reducing their commercial value. Apart from the damaged shell, infestation alters the internal quality of the nut, affecting its flavor and metabolic composition, making it bitter and thus reducing its commercial value [30, 31]. In early infestations, kernel (seed) development is severely inhibited as larvae attack the conductive tissues of the fruit stalk, resulting in malnutrition of the fruit. As a result, these fruits fall off before they reach maturity, or they do not fall off the tree at all and remain on the tree through the winter. Heavily infested fruit can also facilitate the entry of pathogens into the edible interior of the nut, particularly the bacterium *Xanthomonas campestris* pv. *juglandis* (Pierce (Dye)) and the fungus *Marssonina juglandis* (Lib.). This causes the kernel to shrink, lose weight, rot, mold, and deform the kernel, resulting in significant or total yield loss [30–33]. For this reason, the early emergence (mid-July to mid-August) of the walnut husk fly is the most dangerous, whereas the later emergences are not as dangerous because the nuts ripen before the larvae complete their development and cause serious damage. Because the larvae do not complete their development, fewer tannins are released from the husk that would blacken the shell or affect the internal quality of the nuts [30, 31]. Although late infestation usually does not damage the kernel, it interferes with the natural separation of the pulp from the nutshell, making marketing cumbersome and impractical. In addition, black stains must be removed from the shell with high water pressure or the nuts must be bleached because consumers are unwilling to buy stained nuts [4, 32]. In years when vegetation is delayed by cold spring temperatures, the walnut husk fly also reaches its peak attacks later, even in the first days of September. In these cases, even the September emergence of the walnut husk fly causes major damage, especially to late ripening cultivars.

In orchards where walnut husk fly is present, and if left uncontrolled, the damage can be visible on 74 to 91% of the husks [27] and yield loss can be as high as 80% [1]. However, yield losses vary between cultivars.

2. Influence of walnut cultivar on resistance to the walnut husk fly

Some cultivars were found to be more tolerant to walnut husk fly attacks, with first studies reaching in the 1930s, just after the emergence of the pest in California [34]. At the start of the research and with the lack of analyzing equipment that is available nowadays, cultivar resistance was believed to be depended on the hardiness of walnut husk at the time of oviposition activity of the fly [34]. Later on, it was found that walnut husk flies prefer cultivars that produce larger and heavier fruit. Overall, fruit weight was also correlated with the pupal weight and diapause length, as offspring that developed in larger fruit likely accrued fitness advantages over offspring that developed in smaller fruit. Adult fly longevity was reported to be influenced by cultivars and not any particular physical fruit characteristic [35]. Fruit weight has also been documented to influence infestation rates in other Tephritidae species [36, 37]. There are, however, inconsistent and contradicting results between various authors that were comparing the same cultivars across different orchards and years [35, 38]; therefore, no clear conclusion on which cultivar is more or less susceptible to walnut husk fly attack could be drawn. It was, however, suggested that environmental conditions (soil moisture levels, favorable temperatures, etc.) could affect infestation patterns, by simultaneously influencing fly development rate (delaying or accelerating sexual patterns) or walnut growth (delay in the phenophases) [33, 35, 38, 39].

One possibility of cultivar resistance could be the content and composition of phenolic compounds in the husk at the time of infestation. Phenolic compounds are secondary metabolites found in all plant tissues, as they play a major role in physiological processes, growth, and durability of the plant. Their key role in the plant is related to plant defense against biotic and abiotic stresses [40]. In the past, higher total phenolic content was associated with higher stress tolerance in plants [41, 42]. However, in recent years, increasing attention has been focused on individual phenolic compounds and groups as more and more studies [40, 43–45] show that plants respond to pathogens with only selected individual phenolic compounds and groups. It is hypothesized that some individual phenolic compounds and groups respond more rapidly and are better adapted to short-term stress conditions, whereas others require more time to form and are better adapted to long-term stress conditions. In addition, there is a possibility that plant/cultivar resistance is not related to the total phenolic compound content, but rather is due to the reaction time of the plant and the speed with which it recognizes the pathogen and responds quickly to infection containing it while it has not yet spread. Cultivar resistance could be due to the composition of phenolic compounds or to the reaction time of the plant to the infection, as already observed in walnut anthracnose (*Ophiognomonia leptostyla*) [43] and walnut blight (*X. campestris* pv. *juglandis*), where flavonols, flavanols, and naphthoquinones were observed to have the biggest role in walnuts defense mechanisms [40]. Unfortunately, to date, there have been no investigations of this claim, only studies examining the effects of walnut husk fly infestation on kernel quality and its composition [30, 31].

Because cultivar resistance is usually associated with a higher total phenolic content or the phenolic content of certain groups, breeding and selection of new cultivars usually involves breeding cultivars with a higher content of phenolic compounds. Since breeding is a long-term process that can take more than 20 years for walnuts, and farmers cannot tolerate the damage caused by the walnut husk fly for that long duration, the use of pesticides and other technical measures is crucial and necessary. Since walnut husk fly causes up to 80% yield loss, we cannot imagine walnut cultivation without these measures. However, yield losses vary between cultivars.

3. Walnut husk fly control

Governments, agencies, and policymakers in the agricultural sector are constantly faced with the risk of epidemics and pest outbreaks. In the context of global climate change and world trade, there is growing concern about the environmental and economic impacts of non-indigenous invasive species of parasites and pathogens on crops. With globalization and the growing number of trade routes, the problem is increasing every year. Alien insect species have become an increasing problem worldwide due to their significant ecological and economic impacts [1]. According to some estimates, invasive species cause around €19.64 billion in damage and losses each year in Europe alone [46]. Strict regulations and quarantines are in place to prevent these outbreaks, but ultimately the pathogen finds its way across national borders. At the same time, scientific advances and growing public concern about human and environmental health are prompting legislators and policymakers to enact and enforce pest control regulations [1, 4]. Every year, more and more active substances in the plant protection products are regulated or withdrawn from use. And in the event of an outbreak of an invasive pathogen, the effective response can be very limited. To respond effectively to pathogen outbreaks, alternative plant protection methods must be applied. To be successful, accurate data on the pathogen itself (persistence, overwintering, developmental stages, timing of infestation and extent of damage, reproduction, etc.) are required in addition to varietal resistance. Only with all this detailed information can we successfully control the pathogen and its damage [1].

3.1 Monitoring methods

Walnut husk fly control is primally based on successful monitoring of the pathogen to adapt phytosanitary treatments to the occurrence of the pest. Occurrence of the walnut husk fly is usually determined by capturing the adults, but larvae can also be detected. Monitoring of the adult walnut husk fly begins in mid-July. At this time, yellow sticky traps must be suspended in the tree canopy. According to field observations, the efficiency of observations is better when the traps are suspended higher in the tree canopy [1]. There are a number of different traps from different vendors that can be used. Typically, yellow rectangular PVC sticky traps are used that have been shown to be effective in capturing a range of dipterans [27]. The yellow sticky traps are used alone or in combination with attractant, which has been shown to be more effective [27]. Yellow sticky traps were first used in the 1980s in the U.S. and were proven successful in monitoring walnut husk fly populations [47]. Traps are baited with 3 g of ammonium carbonate to increase trap attraction (**Figure 2**) [27].

For an orchard of 1 ha, one to two yellow sticky traps have proven sufficient. It is recommended to place one of the two traps 2 m above the ground and the second 5–6 m above the ground in the canopy. The traps are visually inspected twice a week or every 3 days for the presence of the pest. It is important to remove the trapped flies and re-coat the sticky layer or change the attractant according to the manufacturer's instructions. Normally, the traps are renewed every 3 weeks, while ammonium carbonate is added every week. When the first adult specimens are observed on the traps, it is important to begin the phytosanitary treatments. After the first insecticide application, the pests must be removed from the traps and the traps monitored once a week for further pest occurrence [27, 48]. Monitoring costs associated with walnut husk fly are estimated to be about €75 per hectare, including labor and materials [27, 48].

Larvae can also be detected, but the method is not as reliable and the major initial infestations that cause the most damage are missed because the larvae are already developing in the husk. To detect the larvae, you must visually inspect the

Figure 2.
Two types of traps that are typically in use, PHEROCON® Trécé trap (A), and REBELL® Amarillo trap (B).

fruit surface and locate the damage where the adult females have poked the holes and laid the eggs in the husk. These holes are not easy to see and can be easily missed. Later observation may be easier as the husk begins to blacken from larval feeding activity [1].

3.2 Control methods

3.2.1 Insecticide use

Following the detection of the first walnut husk fly specimens, walnut producers must begin chemical control. Commercially available insecticides containing the active ingredients dimethoate and fenitrothion are considered efficient in controlling the walnut husk fly. Their efficacy is comparable and both showed efficacy when applied at 1500 g of active ingredient per hectare. Apart from the two mentioned, imidacloprid, cyclaniliprole, thiacloprid, zeta-cypermethrin, chlorpyrifos, spinosad, fosmet, and bifenthrin as active ingredients are considered successful in walnut husk fly control. Considering nut prices, average yields, and insecticide application costs, 1 to 2 insecticide treatments are considered economically viable. When pesticides are sprayed at the right time, two applications are considered sufficient to maintain infestations at acceptable levels. However, effective timing of pesticide application can only be achieved with the effective monitoring techniques. A single insecticide application can save about 50% of production in orchards [27]. Current guidelines assume that an attractant (protein bait) is added to insecticide applications to improve their effectiveness. In the canopy, walnut husk flies feed on the protein baits and also ingest the insecticide. Wider nozzles are recommended for easier application because they form larger droplets. Because of the difficulty of applying insecticides to walnuts, it is currently recommended that only 1/3 of the tree canopy from the north to east sides be sprayed with the insecticide and attractant combination. If walnut husk fly is observed again in the yellow sticky traps after the first insecticide application, the insecticide application may be repeated up to two times depending on country guidelines. The last application must be made at least 3 weeks before the walnuts ripen to avoid residues of the insecticide [48]. Because insecticide application to walnut trees is difficult, research is being conducted on different application techniques: special nozzles, drone application, direct trunk injection. One of these is direct trunk injection of abamectin, which has recently been shown to be a viable method of controlling walnut pests [49].

In addition to traditional insecticidal crop protection methods, a walnut husk fly trap has recently become available. It contains an attractant and an insecticide that first attracts the pathogen, and once in the trap, the insecticide kills the pathogen. The lure is hung in the canopy, after the presence of the walnut husk fly is confirmed with the yellow tapes. The traps are a closed system where the insecticide is impregnated on the inside of the lid, while a bag containing the attractant is in the lower container. However, this method is more suitable for gardeners or protection of individual plants, as 50 to 100 traps are needed for a 1-hectare orchard [48].

3.2.2 Non-chemical control methods

With the absence of registered phytosanitary products for walnut husk fly control, more and more alternative methods are being tested. In addition to chemical ones, there are also some non-chemical control methods that are particularly useful in protected areas, urban areas, and areas where the use of insecticides is not appropriate or the use of pesticides is not considered a safe and viable option.

In organic orchards, the use of clay (calcinated kaolinite) has proven effective as it physically protects the fruit. However, even in low-rainfall regions, four to five applications per year are required. In addition, the application is only suitable for small walnut trees, as it is not possible for larger walnut trees [1].

One of the most efficient methods is ground cover under the canopy. This prevents adult specimens from emerging from the ground and thus prevents further oviposition on the husk. However, this method is considered effective only when the entire ground under the tree is covered; otherwise, females from neighboring trees will disperse to these trees [48, 50].

To control walnut husk fly populations, regular removal and burning of infected fallen fruit along with shallow tillage in the spring and fall under tree canopies may be effective. In the fall, shallow tillage of the soil must be done immediately after collecting walnuts. This will destroy the larvae before they can crawl into the soil and pupate. In the spring, it is recommended that the soil be tilled in April to destroy the pupae before the adult emerge. Tillage in the fall must be done 5 to 10 cm deep in the soil and 10 to 15 cm in the spring [48, 50].

3.2.3 Biological control

Among other methods, biotic methods of walnut husk fly control are also an option. Studies have shown that two entomopathogenic fungi are available to control adult specimens of the genus *Rhagoletis*, one is *Beauveria bassiana* and the other is *Metarhizium anisopliae* [51, 52]. The effect of entomopathogenic fungi on larvae and pupae is limited, but they prove effective against adult specimens. To date, no studies of walnut husk fly control with entomopathogenic fungi are known. However, there are some data on the control of cherry fruit fly (*Rhagoletis cerasi*) with the entomopathogenic fungus *Beauveria bassiana*. Application of *Beauveria bassiana* has been shown to reduce cherry fruit damage by 25–30% when applied to larvae and pupae and by up to 65% when applied to walnut husk fly adults foliar [53]. In addition to entomopathogenic fungi, the use of entomopathogenic nematodes of the genus *Steinernema* in *Heterorhabditis* has also been mentioned as a successful biotic agent in the control of various insects [54]. As with entomopathogenic fungi, the effect of entomopathogenic nematodes on walnut husk fly control has not been studied. However, there were some studies on the effect of entomopathogenic nematodes of the genus *Steinernema* on the control of species related to the walnut husk fly, *Rhagoletis indifferens*. The results showed that entomopathogenic nematodes of the genus *Steinernema* were considered successful biotic agents in controlling larvae (up to 80% mortality) and adults (up to 50% mortality), whereas no effect was obtained on pupae [55]. The first optimal time to apply biotic agents against walnut husk fly would be when the walnut husk fly begins to molt from pupae and fly out of the ground (late spring). To determine the proper time to apply biotic agents, the walnut husk fly must be monitored using the yellow sticky panels. The second effective time to apply biotic agents to walnut husk fly would be when the larvae begin to move into the soil (fall) to pupate [1, 56]. Recent work has shown that entomopathogenic nematodes are compatible with insecticides, so combining the two methods could improve efficiency in controlling walnut husk fly [56].

Previously, the possibility of a natural predator, the parasitic wasp *Coptera occidentalis* Muesebeck (Hymenoptera: Diapriidae), was considered. *C. occidentalis* is native to California, USA, and parasitizes the pupae of some species of the genus *Rhagoletis* [57]. *C. occidentalis* has been reported to parasitize the walnut husk fly, along with *Rhagoletis cingulata*, *R. indifferens*, and *Ceratitis capitata*. In the late 1970s, a massive propagation of *C. occidentalis* began in California to control the walnut husk fly. In the 1980s, the parasitoid was released into the wild for the first time. Although *C. occidentalis* has been continuously released for 30 years, its efficacy in controlling walnut husk fly is considered insufficient [58]. This is thought to be due to the particular bionomy of *C. occidentalis*, as it is a parasitic predator that parasitizes pupae in the soil. Due to the insufficient concentration of attractants (kairomones) emitted from the soil by the pupae, the results are not optimistic [54, 59]. However, the species has been introduced in Slovakia as a biotic agent for control of *R. cerasi* [59]. The second parasitic wasp-parasitizing species of the genus *Rhagoletis* is *Diachasmimorpha juglandis* Muesebeck (Hymenoptera: Braconidae). A solitary parasitoid that parasitizes the pupae and larvae of the genus *Rhagoletis*. *Dorcaschema juglandis* finds its prey by sensing the volatile compounds released by infested fruit. However, *D. juglandis* has not been included in the biotic protection program for walnut husk fly control [60]. Close monitoring of native parasitoids in regions where walnuts are grown has been suggested. The species of natural enemies (*Coptera occidentalis*, *Diachasmimorpha juglandis*) of the walnut husk fly could be considered as sufficient alternatives to the use of insecticides in the control of walnut husk fly [54].

4. Conclusions

The walnut husk fly is not a new pest in the United States, but it is fairly new to Europe (1988). Since the introduction of this invasive species, it has spread to almost all walnut-growing areas. Where it is not currently present, it will most likely emerge in the next few years. In the absence of natural predators, it reproduces at a very high rate and causes very high yield losses (80%) in orchards. Monitoring the occurrence of the pest is the most important part of controlling the walnut husk fly. Only with an effective monitoring system, we can apply insecticides at the appropriate time. Since the application of these insecticides is very difficult for adult walnut trees, new methods for insecticide application are being researched (drone application, special nozzles, trunk injection, etc.). As more and more active substances in pesticides are regulated or phased out each year, other control methods are also gaining importance in the pest management. Mechanical control methods are more or less only suitable for walnuts grown in the protected areas, in urban areas and in areas where the use of insecticides is not appropriate or the use of pesticides is not considered a safe and viable option. The use of biotic control agents needs to be further investigated as it may also be an option, especially in organically managed orchards. In addition, the role of phenolic compounds needs further investigation, as little or no research has been conducted. The implications of this research would greatly benefit our understanding of pathogen control, as well as benefit breeders who could easily determine which walnut cultivars have the ability to resist walnut husk fly attacks.

Acknowledgements

This study is a part of programme P4-0013-0481, which is funded by the Slovenian Research Agency (ARRS).

Conflict of interest

The authors declare no conflict of interest.

Author details

Aljaz Medic*, Metka Hudina, Robert Veberic and Anita Solar
Biotechnical Faculty, Department of Agronomy, University of Ljubljana, Ljubljana, Slovenia

*Address all correspondence to: aljaz.medic@bf.uni-lj.si

References

[1] Verheggen F, Verhaeghe A, Giordanengo P, Tassus X, Escobar-Gutiérrez A. Walnut husk fly, *Rhagoletis completa* (Diptera: Tephritidae), invades Europe: Invasion potential and control strategies. Applied Entomology and Zoology. 2017;**52**(1):1-7

[2] Aluja M, Mangan RL. Fruit Fly (Diptera: Tephritidae) host status determination: Critical conceptual, methodological, and regulatory considerations. Annual Reviews of Entomology. 2008;**53**(1):473-502

[3] Mohamadzade Namin S, Rasoulian G. A review of fruit flies of the genus *Rhagoletis* (Diptera, Tephritidae) of Iran and bordering countries, with the key to species. Vestnik Zoologii. 2009;**43**(1):25-30

[4] Aluja M, Guillén L, Rull J, Höhn H, Frey J, Graf B, et al. Is the alpine divide becoming more permeable to biological invasions? – Insights on the invasion and establishment of the walnut husk Fly, *Rhagoletis completa* (Diptera: Tephritidae) in Switzerland. Bulletin of Entomological Research. 2011;**101**(4):451-465

[5] Yee WL, Hernández-Ortiz V, Rull J, Sinclair BJ, Neven LG. Status of *Rhagoletis* (Diptera: Tephritidae) pests in the NAPPO countries. Journal of Economic Entomology. 2014;**107**(1):11-28

[6] Aluja M, Norrbom A. Fruit Flies (Tephritidae): Phylogeny and Evolution of Behavior. Boca Raton, Florida, USA: CRC Press; 1999

[7] Boller E, Prokopy RJ. Bionomics and management of *Rhagoletis*. Annual Review of Entomology. 1976;**21**(1):223-246

[8] Prokopy RJ, Reissig W, Moericke V. Marking pheromones deterring repeated oviposition in *Rhagoletis* flies. Entomologia Experimentalis et Applicata. 1976;**20**(2):170-178

[9] Bush GL. The taxonomy, cytology, and evolution of the genus *Rhagoletis* in North America (Diptera, Tephritidae). Bulletin of the Museum of Comparative Zoology, Harvard University. 1966;**134**: 431-562

[10] Smith JJ, Bush GL. Phylogeny of the subtribe *Carpomyina* (Trypetinae), emphasizing relationships of the genus *Rhagoletis*. In: Fruit Flies (Tephritidae). Boca Raton, Florida, USA: CRC Press; 1999. pp. 205-236

[11] Rull J, Aluja M, Tadeo E, Guillen L, Egan S, Glover M, et al. Distribution, host plant affiliation, phenology, and phylogeny of walnut-infesting *Rhagoletis* flies (Diptera: Tephritidae) in Mexico. Biological Journal of the Linnean Society. 2013;**110**(4):765-779

[12] Boyce A. Bionomics of the walnut husk fly. *Rhagoletis completa*. Hilgardia. 1934;**8**(11):363-579

[13] Yee WL, Goughnour RB. Host plant use by and new host records of apple maggot, western cherry fruit fly, and other *Rhagoletis* species (Diptera: Tephritidae) in western Washington state. The Pan-Pacific Entomologist. 2008;**84**(3):179-193

[14] Cresson ET. A revision of the North American species of fruit flies of the genus *Rhagoletis* (Diptera: Trypetidae). American Entomological Society. 1929;**55**(4):401-414

[15] Chen YH, Opp SB, Berlocher SH, Roderick GK. Are bottlenecks associated with colonization? Genetic diversity and diapause variation of native and introduced *Rhagoletis*

completa populations. Oecologia. 2006;**149**(4):656-667

[16] Foote RH, Blanc FL, Norrbom A. Handbook of the Fruit Flies (Diptera: Tephritidae) of America North of Mexico. Ithaca, NY: Cornell University Press; 2019

[17] Merz B. *Rhagoletis Completa* Cresson und *Rhagoletis Indifferens* Curran, zwei wirtschaftlich bedeutende Nordamerikanische fruchtfliegen, neu fur Europa (Diptera: Tephritidae). Mitteilungen der Schwizerischen Entomologischen. 1991;**64**(1-2):55-57

[18] Duso C. Sull comparsa in Italia di un Tefritide neartico del noce: *Rhagoletis completa* Cresson (Diptera Tephritidae). Bollettino di zoologia agraria e bachicoltura. 1991;**23**(2):203-209

[19] Ciampolini M, Trematerra P. Widespread occurrence of walnut fly (*Rhagoletis completa* Cresson) in northern Italy. Informatore Agrario. 1992;**48**(48):52-56

[20] Seljak G, Zezlina I. Pojav in razsirjenost orehove muhe (*Rhagoletis completa* cresson) v Sloveniji. Zbomik predavanj in referatov s 4. Slovenskega posvetovanja o varstvu rastlin. 1999. pp. 231-238

[21] Ostojić I, Zovko M, Petrović D. First record of walnut husk fly *Rhagoletis completa* (Cresson, 1929) in Bosnia and Herzegovina. Radovi Poljoprivrednog Fakulteta Univerziteta u Sarajevu. 2014;**59**(64 (1)):121-126

[22] Verheggen F, Escobar A, Giordanengo P, Verhaege A. AVIS de l'Agence nationale de sécurité sanitaire de l'alimentation, de l'environnement et du travail relatif à «l'analyse de risques phytosanitaires portant sur *Rhagoletis completa*». Maisons-Alfort Cedex, France: Anses(Agence nationale de sécurité sanitaire de l'alimentation, de l'environnement et du travail); 2014

[23] Barić B, Pajač Živković I, Matošević D, Šubić M, Voigt E, Tóth M. *Rhagoletis completa* (Diptera; Tephritidae) distribution, flight dynamics and influence on walnut kernel quality in the continental Croatia. Poljoprivreda. 2015;**21**:53-58

[24] European and Mediterranean Plant Protection Organization. *Rhagoletis completa*: Diagnostics. OEPP/EPPO Bulletin. 2011;**41**(3):357-362

[25] Ismay JW. Fruit flies of economic significance: Their identification and bionomics. Bulletin of Entomological Research. 1992;**82**(3):433-436

[26] Sarles L, Verhaeghe A, Francis F, Verheggen FJ. Semiochemicals of *Rhagoletis* fruit flies: Potential for integrated pest management. Crop Protection. 2015;**78**:114-118

[27] Duso C, Lago GD. Life cycle, phenology and economic importance of the walnut husk fly *Rhagoletis completa* Cresson (Diptera: Tephritidae) in northern Italy. Annales de la Société entomologique de France. 2006;**42**(2):245-254

[28] Nufio CR, Papaj DR. Host-marking behaviour as a quantitative signal of competition in the walnut fly *Rhagoletis juglandis*. Ecological Entomology. 2004;**29**(3):336-344

[29] Carsten LD, Papaj DR. Effects of reproductive state and host resource experience on mating decisions in a walnut fly. Behavioral Ecology. 2005;**16**(3):528-533

[30] Solar A, Stampar F, Veberic R, Trdan S. How much walnut husk fly

(*Rhagoletis completa* Cresson) affects nut quality of different walnut cultivars? European Journal of Horticultural Science. 2020;**85**(1):63-74

[31] Solar A, Jakopic J, Miklavc J, Stampar F, Veberic R, Trdan S. Walnut husk fly substantially affects sensory attributes and phenolic contents of the kernels in common walnut. Scientia Horticulturae. 2019;**247**:17-26

[32] Hislop RG, Allen WW. Correlation of walnut husk fly activity, larval infestation period, and harvest quality of early-mid-and late-maturing walnut varieties. Walnut Research Reports. 1983. pp. 42-52

[33] Coates WW. Walnut husk fly: Varietal susceptibility and its impact on nut quality. Walnut. Research Reports. 2005. pp. 1-4

[34] Boyce AM. Influence of host resistance and temperature during dormancy upon seasonal history of the walnut husk Fly, *Rhagoletis Completa* cress. Journal of Economic Entomology. 1933;**26**(4):813-819

[35] Guillén L, Aluja M, Rull J, Höhn H, Schwizer T, Samietz J. Influence of walnut cultivar on infestation by *Rhagoletis completa*: Behavioural and management implications. Entomologia Experimentalis et Applicata. 2011;**140**(3):207-217

[36] Papaj DR. Ovarian dynamics in relation to host quality in the walnut-infesting fly, *Rhagoletis juglandis*. Functional Ecology. 2005;**19**(3): 396-404

[37] Messina FJ. Components of host choice by two *Rhagoletis* species (Diptera: Tephritidae) in Utah. Journal of the Kansas Entomological Society. 1990;**63**(1):80-87

[38] Opp S, Zermeno J, Garrick C, Mooney B, Wilkerson C. Timing and susceptibility of walnut cultivars to walnut husk fly attack–3 years of emergence patterns. Walnut Research Reports. 2001:1-3

[39] Coates WW. Walnut husk Fly: Varietal susceptibility and quality observations. Walnut Research Reports. 2004:1-3

[40] Medic A, Jakopic J, Hudina M, Solar A, Veberic R. Identification and quantification of major phenolic constituents in *Juglans regia* L. leaves: Healthy vs. infected leaves with *Xanthomonas campestris* pv. *Juglandis* using HPLC-MS/MS. Journal of King Saud University - Science. 2022;**34**(3):101890

[41] Estiarte M, Filella I, Serra J, Peñuelas J. Effects of nutrient and water stress on leaf phenolic content of peppers and susceptibility to generalist herbivore *Helicoverpa armigera* (Hubner). Oecologia. 1994;**99**(3):387-391

[42] Jiang S, Han S, He D, Cao G, Fang K, Xiao X, et al. The accumulation of phenolic compounds and increased activities of related enzymes contribute to early defense against walnut blight. Physiological and Molecular Plant Pathology. 2019;**108**:101433

[43] Medic A, Solar A, Hudina M, Veberic R. Phenolic response to walnut anthracnose (*Ophiognomonia leptostyla*) infection in different parts of *Juglans regia* husks, using HPLC-MS/MS. Agriculture. 2021;**11**(7):659

[44] Mikulic-Petkovsek M, Slatnar A, Veberic R, Stampar F, Solar A. Phenolic response in green walnut husk after the infection with bacteria Xanthomonas arboricola pv. Juglandis. Physiological and Molecular Plant Pathology. 2011;**76**(3-4):159-165

[45] Solar A, Jakopic J, Veberic R, Stampar F. Correlations between Xanthomonas arboricola pv. Juglandis severity and endogenous juglone and phenolic acids in walnut. Journal of Plant Pathology. 2012;**94**(1):229-235

[46] Haubrock PJ, Turbelin AJ, Cuthbert RN, Novoa A, Taylor NG, Angulo E, et al. Economic costs of invasive alien species across Europe. NeoBiota. 2021;**67**:153-190

[47] Riedl H, Hoying SA. Seasonal patterns of emergence, flight activity and oviposition of the walnut husk Fly in Northern California. Environmental Entomology. 1980;**9**(5):567-571

[48] Solar A. Lupinarji: oreh, leska, kostanj, mandelj. Ljubljana: Kmečki glas; 2019

[49] Kiss M, Hachoumi I, Nagy V, Ladányi M, Gutermuth Á, Szabó Á, et al. Preliminary results about the efficacy of abamectin trunk injection against the walnut husk fly (*Rhagoletis completa*). Journal of Plant Diseases and Protection. 2021;**128**(1):333-338

[50] Olson WH, Buchner RP. Leading edge of plant protection for walnuts. HortTechnology. 2002;**12**(4):615-618

[51] Daniel C, Wyss E. Field applications of *Beauveria bassiana* to control the European cherry fruit Fly *Rhagoletis cerasi*. Journal of Applied Entomology. 2010;**134**:675-681

[52] Yee W, Lacey L. Mortality of different life stages of *Rhagoletis indifferens* (Diptera: Tephritidae) exposed to the entomopathogenic fungus *Metarhizium anisopliae*. Journal of Entomological Science. 2005;**40**:167-177

[53] Daniel C, Wyss E. Susceptibility of different life stages of the European cherry fruit fly, *Rhagoletis cerasi*, to entomopathogenic fungi. Journal of Applied Entomology. 2009;**133**(6):473-483

[54] Laznik Ž, Trdan S. Possibilities of walnuts (Juglans Spp.) protection against walnut husk Fly (*Rhagoletis completa* Cresson) with special emphasis on biological control. Acta agriculturae Slovenica. 2013;**101**:287-292

[55] Yee WL, Lacey LA. Stage-specific mortality of *Rhagoletis indifferens* (Diptera: Tephritidae) exposed to three species of *Steinernema* nematodes. Biological Control. 2003;**27**(3):349-356

[56] Laznik Z, Trdan S. The influence of insecticides on the viability of entomopathogenic nematodes (Rhabditida: Steinernematidae and Heterorhabditidae) under laboratory conditions. Pest Management Science. 2014;**70**(5):784-789

[57] Granchietti A, Sacchetti P, Rosi MC, Belcari A. Fruit fly larval trail acts as a cue in the host location process of the pupal parasitoid *Coptera occidentalis*. Biological Control. 2012;**61**(1):7-14

[58] Nechols JR. Biological Control in the Western United States: Accomplishments and Benefits of Regional Research Project W-84, 1964-1989. Oakland, California, USA: UCANR Publications; 1995

[59] Vallo V. Comparison of laboratory and natural population of *Coptera occidentalis* (Mues.)(hymenoptera, Proctrotrupoidea, Diapriidae). In: Proc. of Symposium, Ecological Problems of Plant Protection and Contemporary Agriculture. Slovakia: The High Tatras Stará Lesná; 1996

[60] Henneman ML, Dyreson EG, Takabayashi J, Raguso RA. Response to walnut olfactory and visual cues by the parasitic wasp *Diachasmimorpha juglandis*. Journal of Chemical Ecology. 2002;**28**(11):2221-2244